TM 508 SIE

DYNAMIC LOADING AND CHARACTERIZATION OF FIBER-REINFORCED COMPOSITES

DYNAMIC LOADING AND CHARACTERIZATION OF FIBER-REINFORCED COMPOSITES

ROBERT L. SIERAKOWSKI
SHIVE K. CHATURVEDI

The Ohio State University

A Wiley-Interscience Publication

JOHN WILEY & SONS, INC.

New York/Chichester/Weinheim/Brisbane/Singapore/Toronto

This text is printed on acid-free paper.

Library of Congress Cataloging in Publication Data:
Sierakowski, R. L.
    Dynamic loading and characterization of fiber-reinforced
composites / Robert L. Sierakowski, Shive K. Chaturvedi.
        p.   cm.
    Includes index.
    ISBN 0-471-13824-X (cloth : alk. paper)
    1. Fibrous composites—Testing.   2. Fiber reinforced plastics–
–Testing.   3. Materials—Dynamic testing.   I. Chaturvedi, Shive K.
II. Title.
TA418.9.C6S52   1997   96-26225
620.1'18—dc20

Printed in the United States of America

10 9 8 7 6 5 4 3 2 1

To

Sandra and Steven
— Robert L. Sierakowski

Chandra and Ashutosh
— Shive K. Chaturvedi

CONTENTS

PREFACE

Currently available textbooks dealing with advanced composite materials examine a number of topics, including the material aspects such as the physical and mechanical metallurgy of such systems, mechanics aspects such as the micro and macro mechanics of such systems, and the mechanics of structural elements composed of composite materials. A paucity of information dealing with the important subject of the dynamic response of composites is available. A number of factors contribute to this lack of knowledge, including the recognized difficulties in assessing dynamic events in monolithic materials with compounded effects extended to advanced composite materials with multifaceted and complex failure modes. It is the purpose of this text to focus on this important topic since one of the key design issues associated with the introduction and extended use of advanced composite materials is related to the dynamic performance of such systems.

The concepts presented in this book are intended to introduce practicing engineers to the important issues associated with this technology, and to gain insight into the current technical literature available for developing an understanding of the behavior of such systems for design purposes. This book is also intended for use in advanced composite graduate curricula. To this end, the subject development progresses systematically by first introducing the types of material systems, followed by a classification of the important type of dynamic loadings considered and dynamic loading regimes. Analytical and experimental modeling approaches associated with local/global response of dynamically loaded advanced composites are also presented. While these introductory remarks are somewhat abbreviated, the important concepts necessary to develop appropriate models as dynamic response predictors are introduced.

Following the first chapter presentation dealing with the concepts and classification of dynamic events, a discussion of the dynamic properties of composites is presented. In particular, the inherent difficulties associated with developing dynamic test procedures is discussed with reference to the inherent interaction between *a priori* knowledge of the properties themselves, the governing constitutive relation for the material requiring knowledge of the properties, and wave propagation effects requiring knowledge of both properties and constitutive relations. A discussion of the various test techniques used to determine dynamic properties of monolithic materials follows, and an extension of these test techniques to advanced composites is discussed. A discussion of currently available dynamic test data as obtained from each of the test techniques described is included.

The next chapter discusses the wave motion in fiber-reinforced composites, with a presentation of a number of the theories that have been advanced to describe wave motion passage. An in-depth examination of the effective modulus theory as applied to composite rod and plate type structural elements is presented. Examination of a composite plate subjected to impact loading is examined, followed by experimental studies conducted on impacted composite plates.

The subsequent chapter examines the ideas associated with the non-destructive testing of advanced composites for use in assessing the damage state in dynamically loaded composite structures. An examination of the status of currently available non-destructive evaluation (NDE) techniques, including their applications and limitations, is discussed. Attention is given to those testing techniques that are most widely used in practice. The remainder of the chapter includes an assessment of glass, graphite, and Kevlar epoxy composite systems with regard to damage modes and their growth during impact loadings and post-impact bending tests. A few important geometrical and material parameters that distinguish the type and extent of damage growth in the above-mentioned composite systems have also been identified.

The chapter on damage tolerance discusses a current and important issue related to damage in composites. A general damage-tolerant design philosophy is examined, as is the effect of damage size and residual strength/stiffness on the performance of damaged composites. Other factors affecting the damage tolerance of composites through residual dynamic response including vibratory and wave propagation effects are also discussed. In addition, the effects of material constituents, fiber-matrix interface, and stacking sequence are discussed.

These ideas are carried forward in the chapter on impact damage modeling in which a classification of model types is presented. Important metrics defining the damage state are presented, as are the important parameters defining the damage event. Several models used for quantifying the damage state are discussed, including an energy balance model and spring-mass model.

Concluding remarks are presented at selected places in the book that summarize the key points and raise issues related to important research needed.

We would like to express our appreciation to the students at the University of Florida and The Ohio State University who have been exposed to and who have provided constructive input to the text. Our special thanks to Aly Emam for his wide variety of help during the final phase of preparing the manuscript; Cindy Sopher for her dedication in careful word processing; Kevin Taylor for the figures; and Gregory Franklin of John Wiley & Sons, Inc. for encouragement and support in the preparation of this manuscript. In addition, we thank the large number of researchers within the professional community who have contributed both directly and indirectly through comments at professional meetings, through publications, and in private conversation to the development of the material presented.

ROBERT L. SIERAKOWSKI
SHIVE K. CHATURVEDI

CHAPTER 1

INTRODUCTION

In reviewing technology advances though the centuries, it is evident that materials development plays a key role in significant technology breakthroughs. If one but reflects on certain historical eras, materials have been either identified with the period or have been critical to resulting developments within the period. Included are the stone age, iron age, industrial revolution, nuclear age, and electronic revolution. Today, with the increasing need for performance-oriented material and structural systems, the development and introduction of advanced composite materials represents a new revolution in materials technology. These new materials represent a marriage of diverse individual constituents which, in combination, produce the potential for performance far exceeding that of the individual elements. This synergism makes composite materials both enabling and pervasive in government and commercial applications.

1.1 TYPES OF COMPOSITES

Advanced fiber-reinforced composite materials are formed by embedding high strength, high stiffness fiber materials within a surrounding matrix of a constituent material. The fibers may be single filaments or multi-filament bundles, the latter being twisted together to form a yarn or tow. The fibers generally used are non-metallic and continuous and are identified as graphite, glass, Kevlar, silicon carbide, boron, or alumina. In addition to continuous fibers, there are also other types of reinforcements such as short fibers, whiskers, platelets, and particulates which are used in discontinuous reinforced composites. Within the types of composite systems discussed, the term *advanced composite* is used to differentiate between those with high performance characteristics—generally strength and stiffness—as opposed to simpler types.

The major classes of structural composites used today consist of polymer–matrix composites (PMC), metal-matrix composites (MMC), ceramic-matrix composites (CMC), carbon-carbon composites (C/C), and hybrid composites. Of these classes of composites, the PMCs are the most widely developed with a wide range of fabricated shapes and accepted commercial properties. These materials are characterized by their light weight, high strength and stiffness, corrosion resistance, and fatigue-resistant properties. Metal-matrix composites are characterized by their higher temperature properties as compared to polymer-matrix composites. Ceramic-matrix composites offer the potential for even higher temperature structural applications when compared to the MMCs. Carbon-carbon composites are superior in applications where very high temperatures occur and where thermal shock is a design factor. Hybrid composites represent the newest class of composites and include the use of a composite material with other composites or with other monolithic materials.

1.2   TYPES OF LOADING

The type of loading applied to the material/structural system can be linked to the time duration the forcing function is applied to the material/structure. The loading function may not necessarily be related to mechanical force but can be represented by a ground displacement, velocity shock, impact, or other loading event. In general, the following classes of loads are recognized as generally applied to material/structural systems:

1. static or dead loading;
2. quasi-static loads applied during material and structural testing;
3. dynamic loads, including

(a) Vibratory

    random

    transient

    steady state

(b) Impact/Impulsive

4. Hydrodynamic loads

In many respects, static and quasi-static loading is linked synonymously with the same testing procedure. Realistically, static refers to very slow, long-term load application while quasi-static is usually associated with generating data from laboratory test equipment such as servo-hydraulic and/or screw-driven test equipment. The loading times associated with these tests are considered to be long enough in duration as compared with the material/structural response such that the internal equilibrium within the material/structure is maintained throughout the loading process. As the loading time is shortened, material/structural inertia effects become important and the loading becomes dynamic. Thus, defining the role of the loading type becomes important in determining the material/structural response. The principal types of dynamic loading in which the system responds as a material/structure, and not as a fluid, can be broadly classified as (1) *vibratory* and (2) *impact/impulsive.*

For vibratory loading, the type of response obtained is directly linked to the applied force. For example, if the forcing function is repetitive and continuous, then the response can be considered as steady state, that is, to a degree, independent of the exact time of application of the forcing function. Sinusoidal forcing functions are often used for representing this type of vibratory response. When the forcing function is non-repetitive and of a finite time duration, the material/structural response is considered to be transient. Once the transient phase has passed, the material/structural response becomes steady state. The system response in the transient stage, however, may result in higher system stresses and displacements when compared to the response in the time regime following cessation of the load. This result is then of concern to the designer. Random vibration effects occur when the instantaneous magnitude of the load is unspecified for any instant of time and the instantaneous magnitudes are specified by probability distribution functions.

The last class of dynamic loads described are associated with impact/impulse type loads. Impact loads are short time loads created by the interaction/collision of two solid bodies, one of which may be at rest. Impulse loads are short time loads produced by striking objects, one of which is not characterized as a solid. For extremely short duration loads, in which the material no longer retains rigidity, the material/structure is said to be exposed to shock loading.

Alternatively to using the load and time of load application as the functional means of classifying the type of loading applied to the material/structure system, the material strain rate can also be considered as a means for identifying the loading type. The linkage between characteristic load times and strain rate effects, as well as methods of loading and dynamic considerations in testing, can be described as shown in Table 1.1.

1.3   DYNAMIC LOADING REGIMES

The types and classes of composites, as previously discussed, are representative of those most widely used in technical applications. In this book, the primary focus is on the dynamic loading and response of such composites. A useful classification schedule for describing the dynamic response of structural elements related to the dynamic loading regimes previously discussed is shown in Table 1.2.

TABLE 1.1   Load Times and Rate Effects

Constant Load or Stress Machine	Hydraulic or Screw Machine	Pneumatic or Mechanical Machines	Mechanical or Explosive Impact	Light-Gas Gun Gun or Explosive-Driven Plate Impact	Usual Method of Loading
10^6–10^4	10^2–10^0	10^2	10^{-4}	10^{-6}–10^{-8}	Characteristic time (s)
10^{-8}–10^{-6}	10^{-4}–10^{-2}	10^0	10^2	10^4–10^6	Strain rate (s^{-1})
Inertia forces neglected			Inertia forces important		Dynamic considerations in testing

In Table 1.2, the pulse duration to structure natural period has been used as a guideline to define the system response. For short-time duration pulses relative to the system natural period, the response can be classified as either an impact event or an impulse event with the loading rise time essentially instantaneous. As previously noted, there is a distinction between the two types of events in that impact involves the collision of two solid bodies while an impulsive loading involves interacting objects, one of which is not characterized as a solid. Material response regimes, on the other hand, can be described in terms of characteristic loading times as identified in Table 1.3.

1.4 PROJECTILE/TARGET CHARACTERISTICS

To define the response of material/structural systems to impact events, the forcing function/intensity time history must be quantified. In order to obtain this information, it is necessary to address the collision event of the respective interacting bodies. In many cases of practical importance, one of the bodies is considered to be initially at rest. For such systems, the body struck is considered as the target while the impacting body is called the projectile. In order to define the forcing function and resulting dynamic events, it is important to

TABLE 1.2   Structural Response Regimes

Regime	Pulse Duration/ Natural Period	Response
1	$t/T < \frac{1}{4}$	Impact/impulse
2	$\frac{1}{4} < t/T < 4$	Vibratory
3	$t/T > 4$	Quasi-static

TABLE 1.3   Material Response Regimes

Regime	Pulse Duration/ Natural Period (s)	Response
1	10^6–10^4	Static
2	10^4–10^2	Quasi-static
3	10^0–10^{-6}	Dynamic
4	10^{-6}–10^{-9}	Hydrodynamic

introduce a definition for the target/projectile interaction. As a simplistic example, the following target/projectile interaction is described:

1. projectile: normal impact;
2. target: initially at rest.

For a given initial velocity and mass of the projectile, a definition of how the target, with specified mass and geometry, may be classified relative to the projectile is necessary. In the broadest sense, target/projectile definitions can be classified as in Table 1.4.

It then becomes extremely important to evaluate the relative degree of softness/hardness of the projectile/target interaction. For example, consider a specific projectile material impacting a stationary target material of specified geometry. One such definition has been advanced by Riera [1], who has introduced a classification parameter β to define the projectile as soft, intermediate, or hard with respect to the target. Specifically, β is defined as

$$\beta = \frac{2(V_y - V_a)C_o}{(V - V_a)^2}$$

where

$$C_o = \text{longitudinal wave velocity in the projectile}$$

$$V = \text{projectile velocity}$$

$$V_a = \text{target velocity}$$

$$V_y = V_a + (\sigma_y/C_o\rho)$$

$$\sigma_y = \text{projectile yield strength}$$

$$\rho = \text{density of projectile}$$

TABLE 1.4   Target/Projectile Definitions

	Target	
Projectile	Hard	Soft
Hard	HS/HT	HS/ST
Soft	SS/HT	SS/ST

HS = hard projectile; HT = hard target; SS = soft projectile; ST = soft target

Thus, a projectile classification relative to the target for the case of a normal impact can be described by Table 1.5.

While the above classification represents a methodology for establishing an impact classification schedule between projectile and target, it does not completely identify the projectile/target interaction and, in addition, considers only the case of normal impact. Although normal impact is considered to be the most unfavorable dynamic loading condition, consideration must be given to the possibility of oblique impact, particularly as it may occur at a critical material/structural location. Depending on the angle of incidence of the projectile with respect to the target, rebound/ricochet can occur. These matters are further compounded by identification of the projectile and the target as being hard or soft. For example, if a soft monolithic projectile interacts with a rigid monolithic semi-infinite target at an oblique incidence angle, shear and bending waves will propagate within the projectile upon contact. These stress waves can produce failure such as rupture of the projectile, thus impeding compressive wave propagation. The force–time history curve of the projectile on the target is thus influenced by the total interactive impact event occurring between projectile and target. While the general problem of impact is complex, it is further compounded by interactions of a monolithic projectile with a composite target, the latter being subject to multi-faceted failure modes.

While the main focus of the preceding discussion has been on the projectile, the role of the target is also of significance in identifying the resulting event. For example, the thickness of the target related to an impact event is an extremely important parameter and can be classified as in Table 1.6.

The degree to which the target is compromised with respect to the projectile can be classified in terms of target penetrability; thus, targets can be considered to be of low resistance, moderately resistant or highly resistant to penetration/perforation. Consider the penetration of a round-ended steel projectile of fixed geometry and velocity striking a semi-infinite target of the materials [2] given in Table 1.7.

TABLE 1.5   Projectile Classification (Normal Impact) [1]*

Class	Soft	Intermediate	Hard
β	$\beta < 0.1$	$0.1 < \beta < 1.0$	$\beta > 1.0$

* Reprinted with permission.

TABLE 1.6   Thickness Classification

Thickness	Defining Property
Semi-infinite	No influence of the distal boundary on the penetration process.
Thick	Influence of the distal boundary only after substantial travel of the projectile into the target.
Intermediate	Rear surface exerts considerable influence on the deformation process during nearly all projectile motion.
Thin	Stress and deformation gradients exist throughout the thickness.

TABLE 1.7   Penetration of Various Materials [2][*]

Material	Penetration (in terms of projectile diameter)
Wet mud	2200
Sand	350
Concrete (2500 lbf/in^2 strength)	36
Concrete (5000 lbf/in^2 strength)	25
Aluminum alloy 2024-T3	1.5
Steel (BHN) 100	0.6
Steel (BHN) 350	0.3

[*] Reprinted with permisison from Elsevier Science Ltd., The Boulevard, Langford Lane, Kidlington OX5 1GB, UK.

It is clear that the degree of penetration is material dependent. In addition, the velocity of the projectile is a most important parameter.

1.5   ANALYTICAL/EXPERIMENTAL MODELING ISSUES

Based upon preceding discussions, it is apparent that in order to develop analytical models and construct meaningful experiments, it is necessary to study a number of research issues. Some of these have been discussed above, such as the type of loading, classification of projectile/target interaction, and local/global response issues. The impact/impulse region is generally identified by stress levels which are two orders of magnitude (plus and minus) of the material yield stress and to particle velocities in the range of the local velocity of sound down to approximately four to five orders below this value.

The general solution approaches to analytically evaluating such problems consist of using

- conservation of mass;
- conservation of momentum;
- mechanical energy balances;
- equivalent static loading (involving prescribed force/velocity conditions).

Related to these matters are such effects as

- defining the impact velocity regimes;
- examining the importance of rate effects;
- describing the damage states;
- establishing the principles of damage tolerance.

Other important parameters can be found in excellent survey articles as described in Backman and Goldsmith [2] and Zukas [3]. The above effects are related in turn to the types of composite materials discussed. In particular, in this book emphasis is focused on the use of advanced composites, exemplified by PMC, MMC, CMC, CC, and Hybrid types with discontinuous and continuous embedded fibers used as the target or projectile materials. The importance of the effects mentioned above, related to the composites studies, are cited in the following paragraphs.

For the case of an impact loading, the response of a material/structural dynamic event can be divided into two general classifications—that of primary and secondary response. The former is associated with effects occurring in the immediate vicinity of the contacting surfaces of the impacting bodies, while the second occurs after the initial impact time and can occur in the far field. The first is governed by stress waves and rate effects while the second is associated with a response governed by the constitutive equations of the material/structure.

For composite targets, the most important element characterizing the response of the structure is the variety and complexity of the damage and subsequent failure modes. Typical dynamic damage associated with impact/impulse events is shown in Figure 1.1.

In addition, the character of the loading itself, the relationship between the magnitude of the load and time, play an important

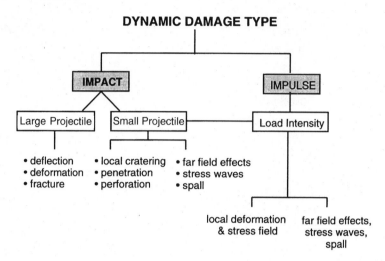

Figure 1.1 Dynamic damage type.

role in the resulting system response. Further, the responses generated are affected by interactions occurring between the impacting bodies and the material structure. Among these factors is the relationship between the projectile/target mass, geometry, rigidity, and impact velocity. The response of composites is affected by the composite constitutive equations, failure of the constituent materials, the interfacial bond between fiber and matrix, and stress wave effects. Both the constitutive equations and the failure modes of the system are recognized to be influenced by rate effects that are in turn dependent upon the magnitude of the impact loading.

In summary, the two principal effects associated with the response to impact loading can be identified as:

1. Contributions of inertia effects of the impacting bodies and the associated wave propagation phenomena, and
2. Changes occurring in the mechanical properties of the system as affected by strain rate effects.

The importance of classifying the impact velocity regimes has been mentioned in a number of the preceding paragraphs. For example, the question as to what constitutes a low velocity, intermediate velocity, and high impact velocity event is dependent to a large extent upon the thickness of the target.

In attempting to quantify the impact event in terms of local and far field stresses as well as resultant damage, for impact regimes where material behavior occurs, the development/availability of an appropriate governing equation is mandatory. As an example of an attempt to quantify an impact effect a characteristic number is presented. This non-dimensional or damage number as given by $\rho v^2/\sigma_y$ can be used to characterize the impact regime for semi-infinite metal targets under normal impact situations. This number is given in Table 1.8 [4].

For the specific case of a mild steel target impacted by a flat-ended steel projectile, it is possible to construct a damage number associated with the table which can be used as a guide to identify impact regimes and the behavior of the projectile/target within the regimes [4]. While the construction of a such a table is useful, some weaknesses can be attached to the use of such a damage number. Among these can be cited the following: non-inclusion of the effect of projectile nose shape, measuring the target yield stress σ_y under intense pressure loading, and the response associated with a predominately elastic loading. Also, the introduction and use of such a damage number has been associated with homogeneous, isotropic materials. The comparable identifying measures for composite materials which can be the target or projectile materials, have yet to be developed. The availability of such measures for classifying and selecting the processes associated with target/projectile interaction for composites should be very useful. It should be noted, however, that it may not be feasible to obtain such identifiers for all classes of composites since the response regimes may be highly dependent upon the type of composite classes. Furthermore, classification may require more than a single parameter identifier.

TABLE 1.8   A Damage Number [4]*

Velocity	$\rho v^2/\sigma_y$	Regime
2.5	10^{-5}	Quasi-static elastic
25	10^{-3}	Plastic behavior starts
250	10^{-1}	Slow projectile speeds
2500	10^{1}	Extensive plastic deformation—ordinary projectile speeds
25000	10^{3}	Hypervelocity impact

ρ = density of target material
v = projectile velocity at normal impact
σ_y = mean flow stress of target material
* Reprinted with permission.

1.6   DYNAMIC THERMAL LOADING EFFECTS

While the above description focuses on the effect of mechanical load-ing, thermal effects are present in all impact phenomena. The degree to which thermal effects retain importance is dependent upon the projectile velocity, material, and the corresponding target thermo-elastic properties and geometry. Since in most cases the impact event is a short-duration event, analyses are usually restricted to wave propagation and impact phenomenon considered as isothermal loading. This type of assumption appears reasonable for slow pro-cesses within the elastic domain of the material/structure. For composites with different constituent elements, heat generated during dynamic deformation and the resulting failure modes are directly linked to the interdependency between local gradient temperature effects and system strengthening/hardening effects. For example, strain hardening effects for MMCs may be important to the resultant failure mode such as catastrophic shear.

1.7   LOCAL/GLOBAL RESPONSE ISSUES

The main focus of this chapter has been on the definition and classi-fication of loading functions associated with impacting bodies rather than impulsive loadings on composite material targets. A further constraint is that one of the interacting bodies, normally the target, is at rest. This latter restriction is introduced mainly for simplicity in classifying one body as at rest and thus denoting this body as the target. It is also generally assumed that in any interactive impact event, in the initial state, all relevant mechanical properties are known. Under this set of circumstances, the following problem types are generally recognized:

1. problems in which information on the question as to whether the target or projectile are damaged during the impact event;
2. problems for which a complete solution associated with the final states of target/projectile are necessary.

The model types introduced for category one may not be sufficient to identify with the second class of problem types and vice versa. Thus, the importance of defining the load time history on the target is also linked to analytical coupling/uncoupling of the projectile/target combination. For example, if one of the interactive bodies is deformed while the other remains relatively undeformed, the problem may be uncoupled in the sense that the behavior of the undeformed body can be classified as a rigid body response. Thus, if the projectile can be idealized as being either soft or, on the other hand, extremely hard (rigid) the problem can be considered uncoupled and the target structure analyzed under the action of a prescribed load time history or displacement time history, respectively.

The importance of classifying the response of a material/structural system is closely linked with identification of the type of loading, that is the load time/reaction time, occurring in the problem. For example, if a dynamic load is transferred to a material/structural system by solid body contact, than depending upon the mass and velocity of the projectile relative to the geometry of the target, the response may be considered either local or global. In addition, if non-linear behavior in and around the impact zone occurs, than a reduction in load transmitted to the rest of the material/structure occurs. Further, if the load is distributed versus concentrated relative to the geometry of the material/structural system the response may be classified as either local or global. Thus, to a large degree, the response is directly linked to the definition of the load occurring. Since focus in this text is on impact loading events, the evaluation and classification of the system response as being local versus global is tied directly to the definition of impact load. To some extent, the expected failure mode also enters into the predictive response, for example, a local punching failure as opposed to large bending deformations. Therefore, it is absolutely necessary to pursue the mechanical response process from the initiation of the impact event up to its termination, and to detemine the mechanical behavior of both bodies involved in the impact event. Finally, if the load application is locally concentrated rather than distributed over a large area, local response effects dominate with global effects negligible; the reverse argument also holds true.

1.8 REFERENCES

1. Riera, Jorge D. (1982) "Basic concepts and load characteristics in impact problems," in *Concrete Structures*, under "Impact and Impulsive Loading," Introductory Report, BAM, Berlin (West), June 2–4, pp. 7–29.

2. Backman, M.E. and Goldsmith, W. (1978) "The mechanics of penetration of projectiles into targets," *Int. J. Engng. Sci.* **16**, 1–99.

3. Zukas, J.A. (1980) "Impact dynamics," in *Engineering Technologies in Aerospace Structures, Design, Structural Dynamics, and Materials,* Editor, J.R. Vinson, The American Society of Mechanical Engineers, New York, pp. 161–198.

4. Johnson, W. (1972) *Impact Strength of Materials,* Edward Arnold, London.

CHAPTER 2

DYNAMIC PROPERTIES

2.1 INTRODUCTION

In dynamic events such as impact phenomena, where the material still remains contiguous, the interaction, that is, collision of bodies, requires not only a knowledge of the propagation of stress waves through the material, but also a constitutive equation for the material. The latter equation represents a mathematical description defining the relationship between stress, strain and the time derivatives of stress and strain. In continuous fiber composites these relationships are generally described by laminate analysis. In certain applications as in the case of metal materials, the influence of temperature may also need to be incorporated into the constitutive equations. For continuous fiber composites, depending upon the selected matrix material such as a polymer base matrix, environmental effects, and in particular moisture effects, may become an important consideration.

In any case, essential to the development of the constitutive equations is information on the dynamic properties of the material system under study. The establishment of these properties for metals is in itself a challenging task; for composites this task is further complicated by the directional dependence of the properties due to the anisotropy of the material system.

There are many difficulties associated with finding the dynamic properties of a material system including: the properties to be determined are functions of the rate of loading; damage initiation or growth may occur during the loading process; a clear distinction must be made between material response and structural response. On this last point we note that the response of a structure depends on its geometry, the point of application of the load and the way in which the material comprising the structure responds. Thus in order to find this material response we must separate, or abstract it, from the overall response of the structure which it constitutes. We must thus design experiments which make this separation, or abstraction, relatively easy.

The overall complexities, as discussed above, thus require study of the interactive nature of dynamic events occurring during the loading process. For example, to determine dynamic properties necessary for modeling dynamic events we must understand stress wave propagation and know the material constitutive equations. To understand wave propagation however, we must know the very same dynamic properties which we are seeking to use as input into the constitutive equations. In summary, in order to understand and quantify dynamic effects, we must synthesize our knowledge of all three aspects: stress wave propagation; constitutive equations; and dynamic properties, as shown in Figure 2.1. In order to predict the general behavior of composites we use experience gained from studies on the dynamic behavior of metals. To this end it is useful to establish a range of strain rate regimes in which experimental test techniques are applicable. Figure 2.2 provides a useful reference for this purpose.

As noted previously, experience associated with the dynamic behavior of metals play an important role in the development and design of experiments for composites. For example, it is generally accepted that rate effects for face-centered cubic (FCC), body-centered cubic (BCC), and hexagonal close packed (HCP) metals usually become important in a strain rate range extending over 50 to 500/s. This strain rate sensitivity for a typical metal is shown in Figure 2.3, where the banded area exhibits the range where rate effects for metals become important. In particular, the experimental test techniques indicated in Figure 2.2 have been found useful for obtaining a number of important dynamic properties such as strength, modulus, fracture toughness, and strain to failure.

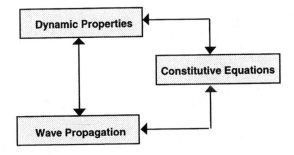

Figure 2.1 Interactive dynamic effects.

Characteristic Loading Times	10^6-10^4	10^4-10^2	10^0-10^{-1}	10^{-2}-10^{-4}	>10^{-4}	(sec)
Strain Rate Regime	<10^{-3}	10^{-3}-10^{-1}	10^0-10^1	10^2-10^4	>10^4	(sec)$^{-1}$
	Constant Load Machines	Hydraulic or Screw Machines	Pneumatic or Mechanical Machines	Mechanical or Explosive Impact	Gas Gun or Explosive Driven Plate Impact Devices	

Figure 2.2 Loading/strain rate regime and experimental test techniques.

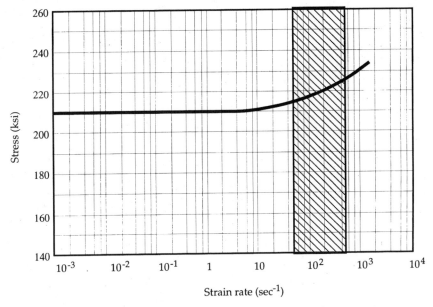

Figure 2.3 Material strain rate sensitivity.

The issues most important in designing experiments for determining the dynamic properties of composites not listed in a rank order are:

1. loading mechanism design to produce the desired stress state;
2. fixture design for holding the specimen in the test assembly;
3. specimen geometry selection;
4. selection of test duration and transit time;
5. measurement of transient parameters;
6. recognition of complex failure modes.

2.2 DYNAMIC TEST METHODS

A number of dynamic test methods are currently used for determining the dynamic properties of composites and are cited in Table 2.1. Each of these test procedures is discussed separately with reference to the materials tested, loading rates, test configurations, and the data obtained.

2.2.1 Punch Tests

- Materials tested
  Weaves, laminates
  (Permaglass, Graphite)
- Punch loading rates
  10^{-7} to 10 m/s
- Test configuration
  beams, plates
- Data obtained

TABLE 2.1   Dynamic Test Methods

Punch tests
Izod, Charpy impact
Drop-weight tests
Hydraulic/pneumatic machines
Hopkinson pressure bar
Flyer plate
Expanding Ring

Maximum punch load vs punch speed
Force–time histories
Fracture damage

Low-speed tests such as those characterized by applied loading at the upper limits of servo-hydraulic and screw type machines rely upon the controlled velocity of the crosshead motion to impart loads to specimens. The punch test is a low-speed test and is generally used for obtaining data on the shear strength of resin matrices and composites.

This type of test is recognized as a way of testing materials and structure for their resistance to penetration and/or perforation under high velocity projectile impacts. There are many situations in which punch type failure must be considered: armament penetration problems; containment of fragments generated in aircraft turbo-transportation engine disintegrations; and, potential hazards to impact by sharp objects to containers carrying nuclear materials. Shieh [1] has reviewed the punch type work related to shipping containers. Much of the work related to such practical situations has dealt with monolithic metals and their structural configurations. In recent years, some attempts have been made to analyze punch failure of modern composite materials, but the processes are not yet fully understood. Thus, we cannot as yet quantify, analyze, and interpret the punch resistance of various composites. It is, however, essential that we can develop analyses for evaluating complex fracture and failure modes that occur during punch type events in composites.

For this type of test the specimen is designed so that it can be rigidly clamped to a stationary fixture; ancillary deflections are not permitted. The loading is applied by a device capable of delivering a constant-speed loading ram. Such a device is used for finding the shear strength of plastics, as shown in Figure 2.4. In preparing test specimens, care must be exercised to ensure that specimen geometry, conditioning, and testing follow accepted procedures (for example, ASTM Standard D732).

Some punch type tests on composite weaves and laminates have been conducted by a number of investigators to obtain data on the maximum punch load versus punch speed, force–time histories and type of fracture and damage expected. Harding [2] has studied the response of composite weaves and laminates for epoxy- and polyester-based matrices. In these tests, which were performed over a punch speed range of 10^{-7}–25 m/s, data on punch speed versus

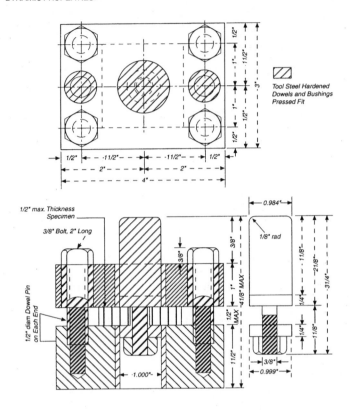

Figure 2.4   A typical punch-test loading device.

load displacement were obtained. Some typical data obtained for an epoxy-based material are shown in Figure 2.5. To be noted is the fact that for epoxy-based material the maximum punch load as a function of punch velocity was found to increase by over 250% as compared to the static maximum load. On the other hand, for polyester matrix materials, the punch load-velocity effect was significant but less dramatic amounting to a 100% increase as compared to the static maximum load.

Some tests performed by Duffey et al. [3] on Kevlar woven panels have been used to compare experimental data obtained for load-deflection curves with a finite element model. It was found that up to the point where complete punch failure occurred, there was a reasonable correlation between the finite element model and experiment in the load deflection response.

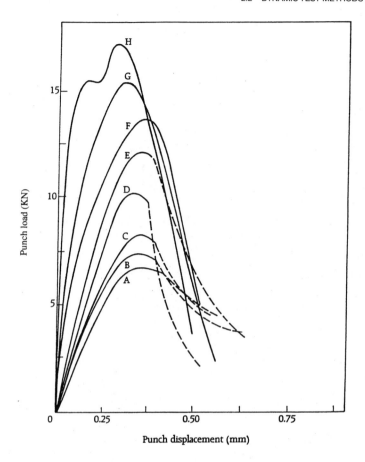

Figure 2.5 Punch tests results for glass-epoxy composites [2]. (Reprinted with permission from Institute of Physics Publishing Ltd.)

2.2.2 Charpy/Izod Tests

- Materials tested
  E-glass/epoxy, S-glass/epoxy, Kevlar 49,
  HMS-Graphite/epoxy, Thornel 75/epoxy,
  Boron/epoxy, Modmore I/epoxy,
  Modmore II/epoxy,
  Boron/aluminum
- Loading rates
  0.005–1000 mm/s
- Test configuration
  Three-point bending tests

· Data obtained
  Energy absorption
  Notch sensitivity
  Fracture behavior
  Rate effects

Charpy/Izod tests have been used for many years to determine the energy absorption, notch sensitivity, fracture toughness and fracture behavior of monolithic materials through information obtained from standardized type pendulum–hammers breaking standardized specimens in a bending mode. For these tests an impact load is produced by swinging a weight from a fixed height into a notched specimen. The notch is introduced into the material specimen in order to produce a stress concentration and thus promote failure in the case of ductile materials. A typical Charpy impact specimen consists of a rectangular cross section beam notched at the beam mid-point, simply supported, and struck by the impacting weight at this point. The Izod specimen is a cantilever beam notched on the tension side of the specimen to insure that fracture occurs under the impact load. Typical specimens for each type of test are shown schematically in Figure 2.6, while typical Charpy/Izod impact pendulums are shown in Figures 2.7(a,b).

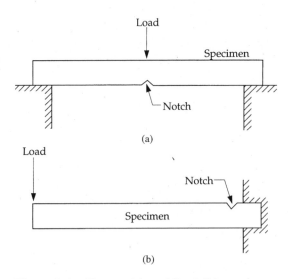

Figure 2.6   Charpy (a) and Izod (b) specimens.

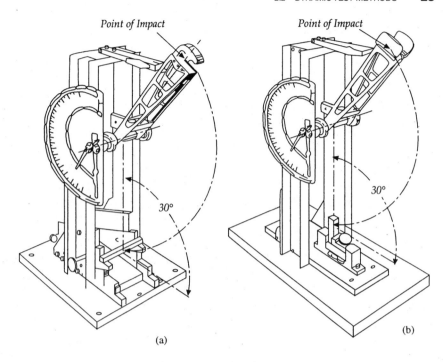

Figure 2.7   Typical Charpy (a) and Izod (b) impact pendulums.

In each case, the impact is produced by swinging a weight W against the test specimen from a height h. When it is released the weight swings through an arc, hits the target specimen and, after fracturing it, reaches a height h'. The difference between the initial energy and the remaining energy of the weight represents a measure of the energy required to fracture the specimen. The energy lost by the pendulum can be expressed as the sum of various energies, these being: the energy required to initiate fracture of the specimen; the energy required to propagate fracture across the specimen, and the energy required to propel the broken specimen. There are other minor sources of energy dissipation: bending of the specimen; vibration of the pendulum components; friction effects within the pendulum system. For brittle materials, and for carefully designed impact devices, it is generally accepted that the first and third of these energy loss mechanisms account for most of the absorbed energy. For composites which are inherently brittle, the complicated fracture process may involve other energy absorption processes. For tough, ductile or fiber-filled materials, the most important factors are the energy required to propagate fracture through the specimen, vibration of the pendulum, and friction at the impact point. For polymeric matrix composites it is also important to consider the environmental conditioning of the specimens prior to the impact event. ASTM D-256 divides impact effects into four categories: a complete break of the specimen, denoted by (C); an incomplete hinge break, denoted by (H); a partial break denoted by (P); and a non-break event labeled (NB).

In addition to understanding the mechanics of the test procedure, we must develop an understanding of the stress state in the specimen during the loading process. Marin [4] recognized that there is a complex stress state in the vicinity of the notch. This notch introduces a size effect for the specimen design and care must be exercised in the interpretation of the test data. Chamis [5] tested Charpy impact specimens following ASTM D-256, for six unidirectional fibrous composite systems. He showed that near the notch, there is a biaxial stress state which is dominated by transverse tensile and interlaminar shear, while near the load application point, transverse compression and interlaminar shear dominate. There are many possible fracture modes: interlaminar shear below the notch root; transverse tension and interlaminar shear at the notch root; transverse compression combined with interlaminar shear and longitudinal compression near the load application point; and, interlaminar shear near the specimen center. These results have been obtained using NASTRAN as the finite element tool assuming linear material behavior and an equivalent static load.

For composites, Charpy/Izod tests have been widely used for obtaining such data as material energy absorption, specimen notch sensitivity, and composite fracture. Initially, such tests have been designed to examine the energy-absorbing qualities of metals and resin matrices. Some of the composite studies using instrumented Charpy tests include, for example, those by Beaumont et al. [6]. In these studies the impact velocities were of the order 3.5 m/s and represented an impact regime where wave propagation effects are ignorable. For such events, typical load time histories are as indicated in Figure 2.8. An important feature of this data is the introduction of a so-called Ductility Index which was introduced by the authors of reference [6] and defined as the ratio of the initiation energy to the propagation energy. This ratio is an assessment of the energy absorption within the material and can be considered as a useful screening tool for the evaluation of the energy absorption of composites.

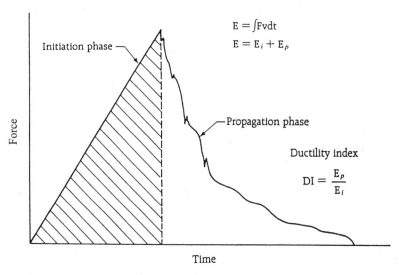

Figure 2.8 Typical load history of impact test [6]. (Copyright ASTM, reprinted with permission.)

Krinke et al. [7] have studied notched and unnotched Charpy test specimens. Figures 2.9(a,b) show some results obtained for composite specimens; these show the effect of strain rate on ultimate bending strength and on specimen energy absorption. These tests provide an insight into the inherent differences between observed static and dynamic failure modes; they also indicate what phenomena can be predicted from analysis and uniaxial test data.

While Charpy/Izod tests have proved to be useful for establishing such information as the time to fracture and the fracture toughness of the material, a number of important questions remain unsettled on the test procedure. For example, it is assumed that in the Charpy/Izod test mode the specimen is in a state of quasi-static equilibrium and that inertial forces can be ignored. For tests in which the time to failure greatly exceeds the time it takes for stress waves to propagate within the specimen, generally acceptable data is obtained. Also, for brittle materials subject to high rates of loading, the energy delivered to the system must be greater than that required to fracture the specimen, and fracture cannot occur at the maximum load. For successful data collection, proper electronic instrumentation is also important. This instrumentation should be selected such that the time to failure exceeds the response time of the instrument by some appropriate measure, for example, at least 10%.

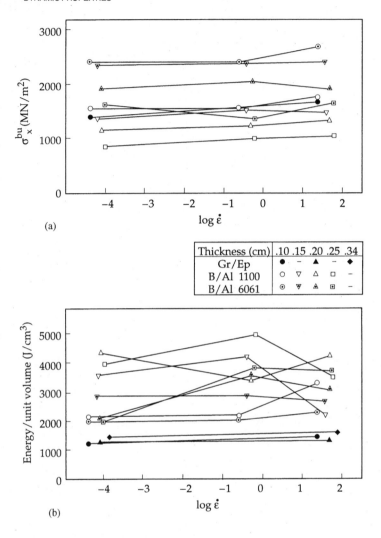

Figure 2.9 (a) Effect of strain rate vs. ultimate bending stress for graphite-epoxy and boron aluminum composites [7]. (b) Effect of strain rate vs. specimen energy absorption for graphite-epoxy and boron-aluminium composites [7].

Ruiz and Mines [8] have suggested an alternative to the Charpy/Izod test to avoid some of the difficulties cited above. They introduce a Hopkinson pressure bar as an alternative loading apparatus. Figure 2.10 depicts an impact bar I, projected with a velocity V_I at a stationary and instrumented receiver bar R, which is already in contact

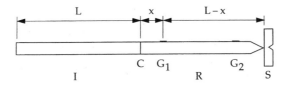

Figure 2.10 Pressure bar schematic [8]. (Reprinted with permission from Kluwer Academic Publishers).

with a notched and fixed specimen S. A compression wave starts at the impact station, between the impact bar and receiver bar, and propagates along the transmitter bar R with a wave speed c equal to $\sqrt{E/\rho}$. Here E is the modulus of the receiver bar and ρ is the material density of the transmitter bar. The wave speed in R is recorded by means of strain gages attached to the bar and located at stations G_1 and G_2.

The strain gages are used to measure the pulse propagation, shape, and amplitude as a function of time. As an example, consider a section of the receiver bar R as shown in Figure 2.11 defined by the two positions A and B on the bar. If the compressive stress pulse amplitude corresponds to a strain $\varepsilon = \sigma/E$, then the pulse traveling with velocity c will reach B at a time equal to $1/c$. At this time, the section AB of the bar will be compressively loaded and Section A will be under compression, with A moving to A′ where the compressive change AA′ is measured by εl. The particle velocity of A can be described by $v = \varepsilon c$ and the corresponding stress intensity given by $\sigma = \rho c v$. Figure 2.12 shows the impactor bar striking the transmitter

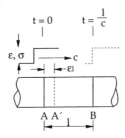

Figure 2.11 Receiver bar geometry [8]. (Reprinted with permission from Kluwer Academic Publishers.)

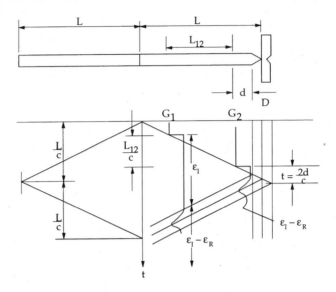

Figure 2.12   Bar geometry and wave propagation diagram [8]. (Reprinted with permission from Kluwer Academic Publishers.)

bar with the specimen fixed in position. For this system, the gages G_1 and G_2 record the passage of the incident pulse ε_I. As the pulse reaches the specimen and reflects back from the contact point C, the reading at gage G_2 consists of both the incident pulse, ε_I, as well as the reflected pulse, ε_R, such that after a time t equal to $2d/c$ the gage G_2 records a combined signal of $(\varepsilon_I + \varepsilon_R)$ while gage G_1 still reads the incident pulse ε_I. A measure of the reflected pulse can be obtained by taking the difference in readings between the signals generated respectively at gages G_1 and G_2. Thus, by storing the signal traces at gages G_1 and G_2 in the transient recorders, displacing the signal at G_1 by the distance L_{12}/c to coincide with G_2, and subtracting the respective signals, we can obtain a measure of the reflected pulse shape ε_R. In addition to obtaining information on ε_R, we can measure the particle velocity $v(t)$ and force $P(t)$ transmitted to the specimen by using the following equations:

$$v(t) = c(\varepsilon_I + \varepsilon_R,) \tag{2.1}$$

and

$$P(t) = AE(\varepsilon_I - \varepsilon_R) \tag{2.2}$$

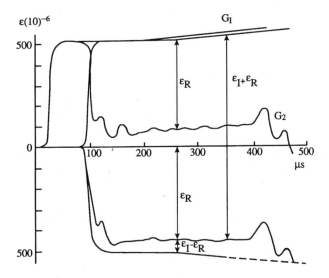

Figure 2.13   Typical strain-time traces [8]. (Reprinted with permission of Kluwer Academic Publishers.)

Some typical strain-time curves, as obtained by Ruiz and Mines [8], are shown in Figure 2.13. These data can, in turn, be used to reconstruct and determine the $v(t)$ and $P(t)$ traces.

There are many advantages of introducing the pressure bar as an experimental tool for determining dynamic properties: the elimination of an irregular strain-time trace as obtained in instrumented Charpy/Izod tests; the ability to obtain load-time information as recorded by the bar strain gages; the ability to determine specimen deflection; the ability to determine the energy absorbed by the specimen; and, confidence in knowing that fracture within the specimen occurs long before the bar stress waves are reflected at the free end of the system.

2.2.3   Drop-Weight Tests

- Materials
  E-glass/epoxy, S-glass/epoxy
  Graphite/epoxy hybrids (ACIF-HT Carbon/Kevlar 49/Araldite F/AT 972)
- Loading rate
  1–10 m/s

- Test configuration
  Beams
- Data obtained
  Energy absorption
  Fracture toughness
  Failure mechanisms
  Strength reduction
  Notch sensitivity

Drop-weight tests have been used instead of Charpy/Izod tests to obtain dynamic properties information on material energy absorption, fracture toughness, failure mechanisms, strength reduction, and notch sensitivity. A schematic of a typical drop weight apparatus is shown in Figure 2.14 [9].

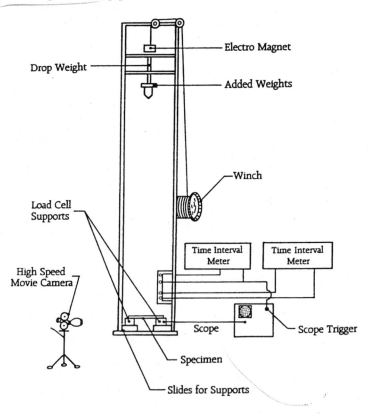

Figure 2.14 Schematic drop impact weight apparatus [9]. (Copyright ASTM, reprinted with permission.)

The apparatus consists of a tower frame with a weight capable of being raised and released with little friction in the vertical direction. The drop weight can incorporate different indenter tips (tups) and the drop weight itself can be varied by adding additional weights. To raise the drop weight a hoist can be used in conjunction with an electromagnet to retain the drop weight at the desired drop height. In a bending impact test, the specimen can be located on supports which themselves can serve as strain gage cells. The span length of the specimen can be adjusted by moving the beam supports. In addition to strain signals which can be monitored on the oscilloscope or digital recorder, a high-speed camera can also be used to obtain real time data on the fracture mode of the specimen. The velocity of the striking drop weight can be measured by means of disruption of photodiode light sources placed at a fixed distance from one another. Upon specimen fracture, two positioned photodiodes below the specimen can be similarly located and used to measure the post-impact velocity of the drop weight. In this way the energy absorbed by fracture of the specimen can be obtained.

Some results obtained for unidirectional and cross-ply E-glass/epoxy composites of different geometries and subjected to different drop heights are shown in Figure 2.15 [9]. Results obtained from these tests suggest that energy absorption is rate sensitive and that the principal mechanism of energy absorption is a delamination type of failure occurring between laminae.

The drop test can also be used to measure both the tensile and compressive properties of composites. In these cases, either information as to the stress-strain curve itself or a particular mechanical property is the object of the test program. In such a device as shown in Figure 2.16, a drop weight is released from a fixed drop distance and impacts the lower grip of a tensile specimen sending a pulse through the specimen. Specimen strain is measured by means of bonded resistance gages attached to the specimen, while the applied load/stress is measured by means of a load cell and accelerometer attached to the upper grip of the specimen.

Tests using such a device have been used on balanced angle ply composites and different fiber-matrix materials as reported by Lifshitz [10]. Some typical stress-strain results obtained for an angle ply laminate are shown in Figure 2.17. In addition to data obtained on composite stress-strain response, dynamic moduli, strength, and data on the failure mechanisms of the particular angle ply configuration

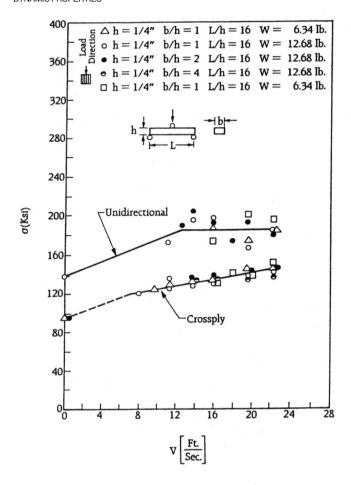

Figure 2.15 E-glass-epoxy strength vs. impact velocity [9]. (Copyright ASTM, reprinted with permission.)

have been obtained. In particular, it has been found that fiber orientation plays an important role in the failure mechanism.

Some drop-weight tests have focused on the strength degradation of composites. Samples tested in bending have been studied by Caprino [11]. In these tests, performed on graphite/epoxy laminates, a two-parameter model based upon linear elastic fracture mechanics concepts has been introduced to predict the residual strength of dynamically loaded composites. The model appears to be a good predictor for relatively low velocity impacts on materials which are rate insensitive. Figure 2.18 shows some typical results obtained by measuring residual strength, σ/σ_0, versus applied energy.

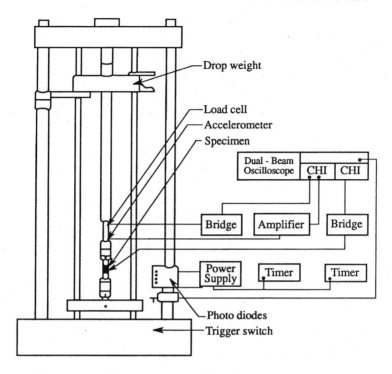

Figure 2.16 Dynamic loading apparatus [10]. (Reprinted with permission from Technomic Publishing Co., Inc.)

Some additional tests on the drop weight response of hybrid loaded composites have been studied by Marom et al. [12]. They have studied the effects of stacking sequence and hybrid structuring of carbon/Kevlar/epoxy composites. Figure 2.19 shows results from impact tests runs with carbon in the center layers of the specimens tested versus Kevlar in the interlayer of the specimens. It is apparent from these data that the impact energy of the Kevlar composites is at least twice that of the carbon. Also to be noted is the fact that as the number of layer interchanges increases, the results stabilize at a point below that predicted by the rule of mixtures.

2.2.4 Concluding Remarks on Charpy/Izod and Drop-Weight Tests

Most of the efforts at characterizing and understanding the impact properties of composite materials have been directed towards the use of Charpy and Izod type tests. These tests, especially their instrumen-

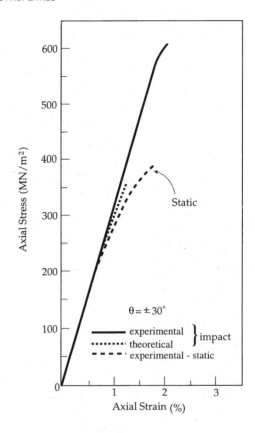

Figure 2.17 Stress–strain curves for ±30° angle ply E-glass-epoxy composite [10]. (Reprinted with permission.)

ted versions, are easy to perform; but it is difficult to relate the results to the impact damage of composites. There are serious fundamental difficulties associated with the observed scatter and variability in data; these arise because there are unknown geometrical and material parameters. For these reasons, such test results can be regarded at best as only semi-empirical.

Composite materials, being highly anisotropic, exhibit many different failure modes: in-plane shear; fiber fracture; delamination; fiber debonding and combinations of these modes. Which failure mode occurs depends upon the specimen thickness and other geometrical size parameters. The design of beam specimens which are used for three-point flexural loading in Charpy tests or under a cantilever type of loading configuration, for Izod tests, appears to be very

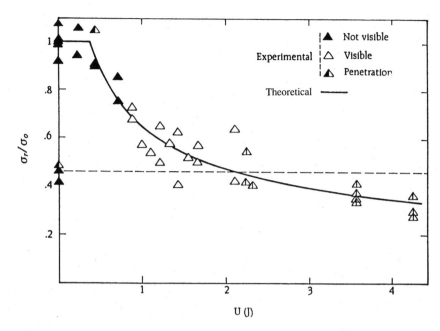

Figure 2.18 Residual tensile strength vs. applied energy, graphite epoxy $(0/\pm 45)_2$ [11]. (Reprinted with permission from Technomic Publishing Co., Inc.)

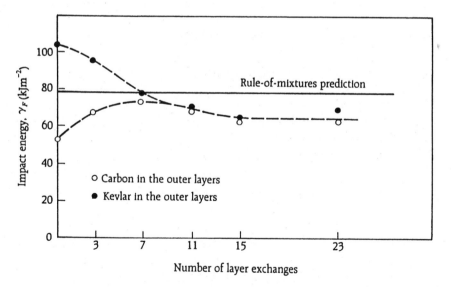

Figure 2.19 Impact-energy related to layer exchange [12]. (Reprinted with permission from Butterworth-Heinemann Journals, Elsevier Science Ltd, The Boulevard, Langford Lane, Kidlington, OX 51GB, UK.)

critical since both normal and shear stresses are developed in these anisotropic beams. The specimen geometry (specimen thickness/ length) ratio can be tailored to produce a pronounced flexural failure in a thin specimen, or a dominant shear failure in a thick specimen. For a laminated composite, a tensile fracture, compressive buckling, or a combination of both types of failure modes is usually observed for flexural failure. This means that it is difficult to determine how energy is absorbed in a Charpy/Izod test on a composite. For a shear type of failure, delaminations (single or multiple) running along the beam length are the more commonly observed fracture modes. Other factors such as loading rate, material nonlinearities, influence of support reactions and complex stress states around the failure zone make it increasingly difficult to predict the impact damage of composites using Charpy and Izod types of tests [13].

Another concern related to these tests stems from situations when the test geometry is not necessarily representative of end-use applications of the composite [14]. In such cases drop-weight tests are recognized to be more relevant, since they represent more closely the plate impact configuration that is typical of many practical applications for composites [15]. The most useful information that can easily be extracted from a drop-weight type of test includes peak force, energy to peak load, total energy to failure and energy losses. Specimen thickness may also play a significant role in impact damage of composites and should be studied. One of the serious problems in the quantitative evaluation of such data is related to the existing oscillations that are observed in load-time plots. These oscillations are recognized to occur primarily due to the falling weight-transducer-target interactive wave propagation and associated vibratory phenomena. Several precautionary measures for data acquisition and interpretation have been suggested in the literature [15, 16] and can be adopted to obtain a reasonable estimate of the impact damage of composites. It should be borne in mind, however, that data generated using different impact test methods—Izod, Charpy, and drop weight—may not correlate with each other, according to a recent comparative assessment of test methods by Kakarala and Roche [17]. In general mathematical models are not available to analyze the impact response of composites, and most of the semi-empirical approaches provide only guidelines to generate and interpret test data. Therefore, selection of composite type and test methods should conform to the stress states, geometric and boundary constraints, and other relevant variable limits as noted for a particular application.

2.3 HYDRAULIC/PNEUMATIC MACHINES

- Materials
  Neat resins (PMMA, CAB, Nylon, Polypropylene)
  S-glass/epoxy (828, 871)
- Loading rates
  1 to 50/s
- Test configuration
  Tension Specimen
- Data obtained
  Strain rate sensitivity
  Failure modes
  Mechanical properties
  Constitutive equation modeling

Hydraulic/pneumatic tests have been used to obtain data on composite strain rate sensitivity, failure modes, dynamic material properties and constitutive equation modeling. Tests of this type are especially useful for controlled strain-rate testing in the medium strain-rate range. Chou et al. [18] studied the room temperature dynamic compressive behavior of a number of neat resins and their associated use for temperatures over a strain-rate range extending from 10^{-4}/s to 10^3/s. A summary of work done on a variety of plastics has also been included by these authors and is presented in Table 2.2. In particular, a unique open/closed loop testing machine has been designed for use in these experiments. The closed-loop mode incorporates a feedback system with a function generator for monitoring load and displacement. The open-loop system uses fast acting valves with various orifice sizes, which when coupled with adjustable piston strokes insures a controlled displacement rate. Typical true stress-strain curves obtained for a PMMA resin are shown in Figure 2.20, as well as true stress and strain rate behavior shown in Figure 2.21.

In the same time period Armenakas and Sciammarella [19] performed dynamic tests using a specially designed high-speed apparatus in which an explosive type launch system was used.

Specimens tested were plate-like fabricated from glass/epoxy with measurements made by high-speed photos taken of Moiré patterns using a Beckman–Whitley camera. Stress–strain rate data for the

TABLE 2.2 Summary of Work Done in Compression of Plastics [18]*

Investigator	Date	Materials	Strain Rates (s^{-1})	Temperature (K)
Kolsky	1949	Polyethylene	1900-7500	293
		PMMA	575	292
		Natural rubber	5000	290
Back, Campbell	1957	Phenol formaldehyde	500	Room
		Phenol formaldehyde composite	500	
Ripperger	1958	Polyethene	2200; 5700	21; 300; 340
		PTFE	2200	300
		Nylon	2200	300
		CAB (Tenite M)	2200	300
		CAB (Tenite H)	2200	300
		Polystyrene	2200	300
		Polyvinylidene chloride	2200	300
Volterra, Baron	1958	Natural rubber	41	274; 294
		Hard rubber	25	
		Polychloroprene (Neoprene GNA)	33	
	1963	Polyethylene	0.4–16	
		Ethyl cellulose	0.4–7.4	233–323
		CAB	0.4–4.3	
Tardif, Marquis	1963	PMMA	150–590	Room
		Polycarbonate	327–1028	
		Nylon	500; 860	
		Polyacetal	450; 670	
		PTFE	570–1, 420	
		Epoxy	150–690	
Davies, Hunter	1963	PMMA	1500; 5600	Room
		Nylon 6	2000; 8000	
		Polyethylene	1270; 3700	
		PTFE	1430	
		PVC	1250	
Lindholm	1964	Epoxy	930	Room
Hoge	1965	Polypropylene	0.6–1610	Room
Maiden, Green	1966	PMMA	0.005–1210	Room
Hold	1968	PMMA	10^{-5}–1000	273–388
Dao	1969	Polyethylene Nylon 6–6	0.0003–33	77–450
Meikle	1969	PMMA	0.07–33	200–360

* Reprinted with permission from the Society for Experimental Mechanics, Inc.

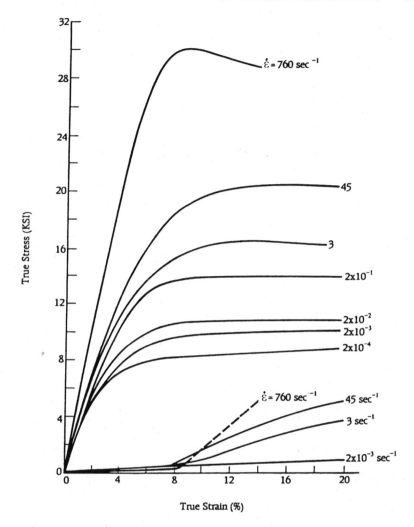

Figure 2.20 Stress–strain curves for PMMA [18]. (Information reprinted courtesy of the Society for Experimental Mechanics, Inc.)

glass/epoxy composite tested is shown in Figure 2.22. Some of the interesting observations made concerning composite failure include the fact that the number of fiber breaks at high rates is higher than that at low rates, and that the strain corresponding to the ultimate load is dependent on the strain at the ultimate load of the individual glass fibers.

Another interesting hydraulic/pneumatic device has been discussed by Matera, and Albertini [20]. This hydraulic/pneumatic device is

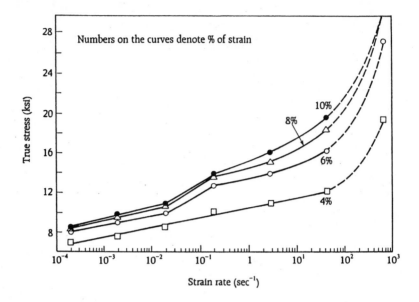

Figure 2.21 True stress vs. strain-rate curves for PMMA [18]. (Information reprinted courtesy of the Society for Experimental mechanics, Inc.)

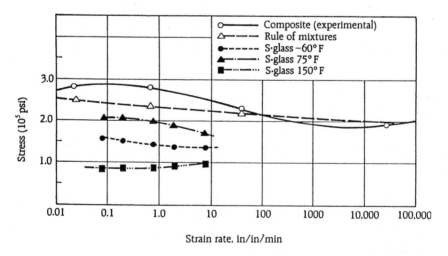

Figure 2.22 Stress vs. strain-rate for S-glass-epoxy [19]. (Information reprinted courtesy of the Society for Experimental mechanics, Inc.)

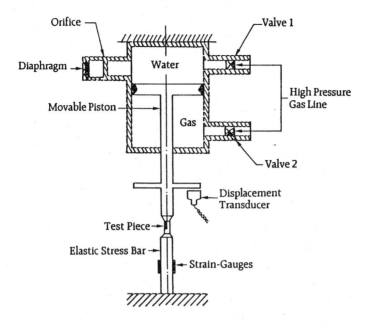

Figure 2.23 Pneumatic hydraulic testing device [20]. (Reprinted with permission from Elsevier Science S.A.)

shown in Figure 2.23 and uses gas and water, the latter placed in the upper chamber, while both chambers are initially at equal pressure. When the diaphragm shown in Figure 2.23 is ruptured, liquid is ejected, the piston moves up, with the pressure controlling the displacement rate. Strain measurements are made by means of a displacement transducer which is fixed to the upper bar which moves with the piston. This device allows for strain rate measurements in the range 1–50/s.

2.4 HOPKINSON PRESSURE BAR (COMPRESSION)

- Materials
  Epoxy C 124, steel/epoxy (Epon 815, 828),
  E-glass/(Epon 815, 828), CuW, Al-Al$_3$Ni, Ni-Nb-Al Du/W,
  St/W, W/Al, W/steel

- Loading rates
  $10^2-10^4/s$
- Test configuration
  Compression specimen
- Data obtained
  Strain sensitivity
  Stress pulse shaping
  Constituent properties
  (filament size, volume fraction)
  Dynamic yield/ultimate stress
  Fracture mechanisms
  Constitutive equation modeling
  Damage initiation

2.4.1   Introduction

One of the most widely used tests for evaluating high strain-rate effects in materials is the Hopkinson pressure bar test. This type of test procedure has been used to examine the dynamic response of materials in various modes of testing. Some of the important data obtained by using pressure bar testing include strain-rate sensitivity, material properties, dynamic yield stress, damage propagation, and fracture/failure mechanisms. Included among the principal pressure bar testing modes used to study rate sensitivity are compression, tension, and shear.

2.4.2   Historical Comments

Hopkinson [21] developed the original technique used to obtain measures of the time duration of an impact load and the maximum pressure levels associated with the impact event. As originally conceived, the apparatus consisted of a bar several feet long and approximately one inch in diameter suspended by threads, with an attached end piece that acted as a momentum trap. The end piece was held onto the main bar by magnetic attraction. A bullet was then fired to impact axially the end of the main bar opposite from the momentum trap. When the compressive wave, reflected as a tensile wave from the free end of the momentum trap, reached the interface between the two bars, the tensile wave broke the weak magnetic attachment and permitted the end piece, which now contained all the momentum, to fly off and leave the main bar in place. The end piece swung upward, supported by threads, and the magnitude of its swing was related to the amount of momentum trapped in it at the beginning of the swing.

This basic pressure bar technique led to a further development by Kolsky [22] who introduced the so-called split Hopkinson pressure bar for obtaining dynamic stress–strain data at room temperature. In the Kolsky system a short compression specimen placed between two elastic pressure bars was tested. Since its introduction, many investigators have contributed to extending and refining this technique. For example, for compression testing of materials at room temperature, Davies and Hunter [23] have discussed the effect of specimen size as related to end friction effects and radial inertia corrections. Further dynamic compressive tests on polycrystalline and single crystal materials using the split Hopkinson bar have been performed by Lindholm and Yeakley [24] and by Larsen et al. [25]. The limitations of using a one-dimensional theory to account for loading and unloading waves in a split bar system have been discussed by Conn [26]. Additional discussions on the use of the Hopkinson pressure bar and the associated one-dimensional equations used in its analysis have been given by Hauser [27]. More recently, a two-dimensional numerical analysis of the governing equations has been presented by Bertholf and Karnes [28]. Their analysis has been used to re-examine radial inertia and end friction effects between the specimen and the elastic bars as well as the effect of the geometrical length-to-diameter ratio of the test specimen, thus reconfirming the conclusions of Davies and Hunter [23]. A method for describing the stress-wave propagation in a split Hopkinson bar using a viscoelastic specimen has been discussed by Chin and Neubert [29]. This technique is based upon the use of a finite-difference scheme for solving the governing wave equations. The relative merits of using surface strain measurements on the specimen as opposed to average strain values in the split Hopkinson bar system have been discussed by Bell [30].

2.4.3   Compression Bar Test Description

For the impactor assembly shown in Figure 2.24 the striker bar is accelerated by means of a driving force supplied from a torsion bar spring mechanism or a gas gun launch system. If a mechanical launch scheme, such as a torsional spring is used, the striker bar is first positioned in a yoke assembly and drawn back by means of a hydrau-

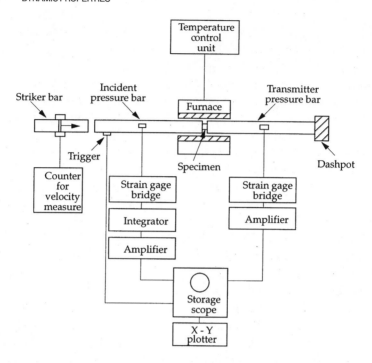

Figure 2.24   Schematic of test assembly.

lic piston connected to the yoke by a shear pin. When the pin shears, the torsion spring propels the striker bar to impact the incident bar axially. The striker bar and the pressure bars are selected to be of the same material and diameter. Thus, upon impact a pressure pulse of approximately constant amplitude and of finite duration is obtained. The striker bar is unloaded from the incident pressure bar when the compression pulse traveling through the striker bar reflects at the free surface as a tensile pulse and returns to the impact face. Therefore, the pulse in the incident pressure bar is twice the length of the striker bar. The amplitude of this pulse is controlled by adjusting the release position of the mechanical spring-loaded mechanism or pressure in a gas launch system. When the pulse reaches an interface along the bar assembly, such as the specimen, a part of the incident pulse is transmitted and a part is reflected. The relative magnitudes of these pulses is dependent to a great extent upon the physical properties of the specimen. Because of the numerous internal reflections, the stress distribution along the specimen is smoothed out and the stress can be considered uniform along the specimen length except for the initial and falling part of the pulse. In order to analyze pressure bar data, the strain–time histories of the incident, transmitted, and reflected pulses need to be recorded. For this purpose strain gages are positioned so that the incident and reflected pulse in the incident pressure bar and the transmitted pulse in the transmitter bar can be recorded without significant distortion. The desired position for locating these gages is documented through use of a Lagrangian x–t wave diagram shown in Figure 2.25 for high-strength tool steel bars. The diagrams show the incident pulse $\varepsilon_I^{(t)}$, the reflected pulse $\varepsilon_R^{(t)}$ and the transmitted pulse $\varepsilon_T^{(t)}$ as well as the dwell time between the passage of the incident pulse and the arrival of the reflected pulse at the gage station on the incident pressure bar. The dwell time is calculated to be 110 μs for the tool-steel arrangement shown. The pulse length Δt_0 in each case is given by $\Delta t = 2L_s/C_0$, where L_s is the striker-bar length and C_0 is the bar-wave speed. The design selection of elements for the pressure bar are summarized in Figure 2.26.

2.4.4 Specimen Size and Friction Effects

For split Hopkinson pressure bar tests, it is generally accepted that a length-to-diameter ratio of one-half is desired in order to minimize

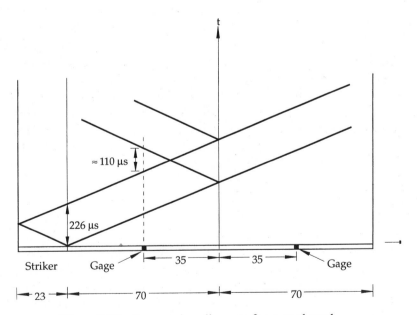

Figure 2.25 Lagrangian diagram for a tool steel.

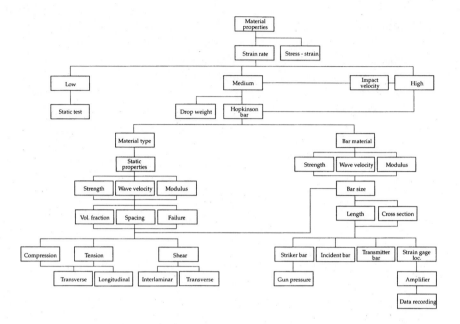

Figure 2.26   Pressure bar design chart.

radial inertia effects. It is also generally acknowledged that the specimen diameter used, while maintaining the above ratio, should be less than the pressure bar diameter. Careful attention is also required in the preparation of the end surfaces of the specimens and their lubrication in order to minimize friction effects. Both radial inertia and friction effects can be examined by using various sized specimens. Specimens can be tested with unlubricated end surfaces and with lubricated (boron nitride coated) end surfaces at room temperature. In most cases, unlubricated specimens produce data indicating slightly higher stresses than that of unlubricated specimens. In addition, barreling of a specimen can sometimes occur.

2.4.5   Derivation of Stress–Strain Equations

In using the split Hopkinson pressure bar, the strain-gage transducers mounted on the incident and transmitter pressure bars are used as signal monitors to determine the strain-time history of the incident, reflected and transmitted pulses in the elastic pressure-bar system. These signals are in turn related to the displacements occurring at the interfaces of the pressure-bar faces in contact with the specimens.

From the interface displacement–time records the average specimen strain can be determined. From elastic bar-wave theory for a pulse propagating into a uniform bar which is initially unstrained and at rest before the pulse arrives, the strain ε at any point is given by

$$\varepsilon = -\frac{v}{c_0} = -\frac{1}{c_0}\frac{du}{dt} \tag{2.3}$$

where $v = du/dt$ is the particle velocity (positive away from the striker bar) and c_0 is the elastic bar-wave speed assumed constant for uniform bar properties and at ambient temperature. Integration of the above equation gives the displacement u as

$$u = -c\int_0^t \varepsilon \, dt \tag{2.4}$$

The striker bar initiates a compressive loading pulse in the incident bar which is partially transmitted through the specimen and partially reflected from the interface. The relative magnitudes of the incident ε_I, reflected ε_R and transmitted ε_T strain pulses depend upon the physical properties of the specimen. Because of the short length of the specimens, many internal reflections can occur in the specimen during the time duration of the loading pulse, since the loading pulse is long compared to the wave transit time through the specimen. Therefore, it can be assumed that the stress is approximately uniform along the short specimen except during the rapidly rising and falling parts of the pulse. Thus, the stress in the specimen can be determined from measurement of the transmitted elastic strain ε_T at the interface of the specimen.

The displacement functions at the incident and transmitter bar interfaces of the specimen can be written as

$$u_I(t) = -c_0\int_0^t [\varepsilon_I(t') - \varepsilon_R(t')] \, dt' \tag{2.5}$$

$$u_T(t) = -c_0\int_0^t \varepsilon_T(t') \, dt' \tag{2.6}$$

The negative signs denote positive displacements (away from striker bar) for negative (compressive) strains ε_T and $(\varepsilon_I - \varepsilon_R)$. The nominal strain rate $\dot{\varepsilon}_S$ in the specimen of initial length L_0 is thus given by

$$\dot{\varepsilon}_s = -\frac{1}{L_0}(\dot{u}_I - \dot{u}_T) = \frac{c_0}{L_0}[\varepsilon_I - \varepsilon_T] \tag{2.7}$$

The corresponding forces at the ends of the pressure bars in contact with the specimen (considered negative for compression) are given by

$$F_I = EA[\varepsilon_I + \varepsilon_R] \quad \text{and} \quad F_T = EA\varepsilon_T \tag{2.8}$$

In the above equations E is the elastic modulus, considered constant for a given temperature, and A is the pressure bar cross-sectional area. For an essentially uniform stress in the specimen, it follows that

$$\varepsilon_T = \varepsilon_I + \varepsilon_R \tag{2.9}$$

and thus the strain rate in the specimen can be written as

$$\varepsilon_S = -\frac{2c_0}{L_0}\varepsilon_R \tag{2.10}$$

The specimen strain can then be found by integrating the above equation. Thus,

$$\varepsilon_S(t) = -2\int_0^t \frac{c_0}{L_0}\varepsilon_R(t')\,dt' \tag{2.11}$$

The above integration is performed by means of an operational integrator/amplifier as shown in the schematic diagram of Figure 2.24. The integrated specimen strain signal $\dot{\varepsilon}_s$ and the corresponding transmitted strain signal ε_T, which is proportional to the specimen stress, can then be fed into the calibrated oscilloscope to produce a dynamic stress-strain record of the test specimen. These data can be graphically displayed using an X–Y plotter attached to an oscilloscope system or stored in a data acquisition system for future use.

2.4.6 Pressure Bar Calibration

Calibration of stress and strain signals requires an accurate measurement of the striker bar velocity. The striker bar velocity can be obtained by several methods and for a mechanical spring launch system a curve of striker bar velocity vs draw back distance can be obtained. This curve is generally a straight line and after being obtained once, need only be checked occasionally. If a gas gun launch system is used then the velocity of the striker bar can be measured by means of breaking photovoltaic light sources located at the end of the launch barrel.

For calibration purposes the two transmitter bars are wrung together without a specimen sandwiched between them. If strain pulses from an impact at known velocity are then recorded at the incident and transmitter bar gage stations, the two recorded pulses should be approximately identical and rectangular in shape as shown in Figure 2.27, which is a tracing of an actual x–y plot recorded from the memory of a digital oscilloscope.

The average strain absolute magnitude $|\bar{\varepsilon}|$ and pulse length Δt are shown in Figure 2.28, which represents schematically a typical incident bar pulse. These values can be calculated from the elastic bar wave propagation conditions in the incident and striker bar. Immediately after the impact at speed v_0 compressive pulses propagate

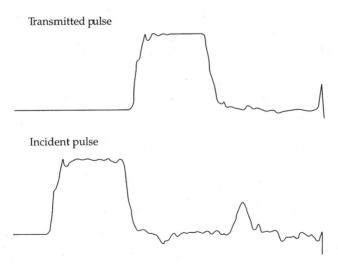

Figure 2.27 Pressure bar pulse shapes with bars wrung together without a specimen.

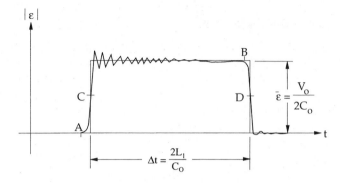

Figure 2.28 Schematic of pressure pulse in incident bar.

away from the impact point in both bars. In the regions adjacent to the impact point the particle velocity is $v_0/2$ and the strain magnitude $\bar{\varepsilon}$ is given by

$$\bar{\varepsilon} = \frac{v_0}{2c_0} \tag{2.12}$$

where c_0 is the elastic bar-wave speed. The pulse duration is

$$\Delta t = \frac{2L_1}{c_0} \tag{2.13}$$

where L_1 is the length of the striker bar. If the pulse is approximated as rectangular then the integral of the incident pulse is $\bar{\varepsilon} \, \Delta t$.

The oscillations or ringing on the top of the pulses of Figures 2.27 and 2.28 are caused by radial (Pochhammer-Chree) oscillations. The irregularities at about the mid-point of the rise time in Figure 2.27 are caused by reflections from the joint where the two 35-inch segments are joined to form the pressure bar. They do not appear in oscilloscope records for a continuous bar.

For calibration of the integrator used to record specimen strain, the integral of the incident pulse is used. This integral may be estimated as $\bar{\varepsilon} \, \Delta t$, using the value of $\bar{\varepsilon}$ given by Eq. (2.12) and Δt given by Eq. (2.13). The time Δt actually corresponds to the time between points A and B in Figure 2.28 but if the same time interval Δt is used between points C and D for the rectangular approximation of the pulse, the area under the curve is seen to be approximated by the area inside the rectangle. With digital recording techniques it is possible to integrate the actual pulse numerically. Alternatively, a shorter time $\Delta t'$ could be used corresponding to an interval where the ε record is nearly constant, and $\bar{\varepsilon} \, \Delta t'$ then compared with the voltage change for the same time interval of the integrated output (on the straight part of the descending curve in the lower trace C of Figure 2.29). Figure 2.29(a) shows a typical recorded stress–strain curve as it appears on the oscilloscope face. The vertical displacement is proportional to the transmitted strain and therefore to the specimen stress. Thus, using Eq. (2.11), we can write,

$$\sigma_s = \frac{EA_B}{A_S} \, \varepsilon_T = -\frac{A_B}{A_S} f_e(\Delta V)_T \qquad (2.14)$$

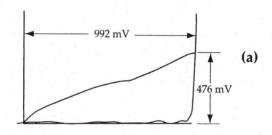

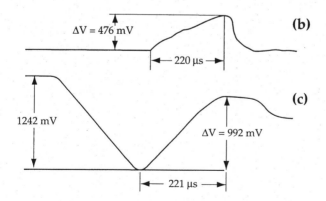

Figure 2.29   Test data for OFHC copper at 25°C: (a) x–y plot representing stress–strain curve of specimen; (b) transmitter bar strain vs. time; (c) integrated strain vs. time in incident pressure bar.

where $(\Delta V)_T$ is the positive voltage indicated as a function of time by the upward vertical displacement of the oscilloscope trace, corresponding to a negative (that is, compressive) strain ε_T, A_B and A_S are the cross-sectional areas of the bar and specimen, respectively, and f_e is a calibration factor. The horizontal displacement of the oscilloscope trace measures the output of the integrating operational amplifier into which the incident and reflected pulses are fed. With proper gating of the oscilloscope trace, the beam remains at point "0" until the incident pulse arrives and then is swept to B by the integral of the incident strain-time signal, establishing a base line of zero stress. Then when the reflected and transmitted signals arrive at the same time the beam is driven back to the right by the integral of the reflected pulse (opposite in sign to the incident pulse) and upward by the transmitted pulse as discussed above. The displacement $(\Delta V)_R$ in volts is related to the specimen strain. Thus, by Eq. (2.11),

$$\varepsilon_S(t) = -2 \int_0^{\Delta t} \frac{c_0}{L_0}\, \varepsilon_R(t')\, dt'$$
$$= -f_e'(\Delta V)_R$$

(2.15)

where f_e' represents a calibration factor.

The two calibration factors f_e and f_e' are determined by omitting the specimen and wringing the two bars together. In this case there is no reflected pulse, with both the incident and transmitted pulses being essentially rectangular pulses of known absolute magnitude $\bar{\varepsilon}$ and duration Δt given by Eqs. (2.12) and (2.13). The oscilloscope records for an actual calibration are shown in Figure 2.30.

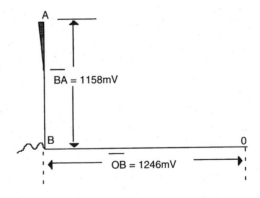

Figure 2.30 Calibration trace for pressure bars.

As the incident pulse $\varepsilon_I = -\bar{\varepsilon}$ passes the incident bar gage station, the integrated output from this pulse moves the beam from O to B in Figure 2.30 just as it does during a test with a specimen in place. When however, this pulse arrives at the transmitter bar gage station there is no reflected pulse to move the trace back to the right while the transmitted pulse moves it upward. Instead the trace moves straight up along BA. With $\varepsilon_T = -\bar{\varepsilon}$ and $(\Delta V)_T = \overline{BA}$, Eq. (2.14) becomes

$$\frac{EA_B}{A_S}\bar{\varepsilon} = \frac{A_B}{A_S}f_e\overline{BA}$$

or

$$f_e = \frac{E\bar{\varepsilon}}{\overline{BA}} = \frac{v_0}{2c_0}\frac{E}{\overline{BA}} \tag{2.16}$$

where the last form is obtained by using Eq. (2.12).

The calibration factor f_e' is determined from the absolute magnitude $\overline{OB}$ in Figure 2.30 by substituting into Eq. (2.15) $\overline{OB}$ in place of $(\Delta V)_R$ and $\bar{\varepsilon}$ in place of $-\varepsilon_R$. This gives

$$2\frac{c_0}{L_0}\bar{\varepsilon}\,\Delta t = f_e'\overline{OB} \tag{2.17}$$

which, by use of Eqs. (2.12) and (2.13), gives

$$f_e' = \frac{2L_1 v_0}{L_0 c_0 \overline{OB}} \tag{2.18}$$

When recording the stress-strain curves in Figure 2.29, the gated time must be adjusted to prevent interference from reflected and retransmitted signals in the bars.

2.4.7   Temperature Corrections for Pressure Bars

Perzyna [31] has analyzed longitudinal elastic-plastic wave propagation in a bar whose properties vary with axial position. For the elastic case the governing equation may be written as

$$\frac{\partial\sigma}{\partial x} = \rho(x)\frac{\partial y}{\partial t}, \qquad \frac{\partial\varepsilon}{\partial t} = \frac{\partial v}{\partial x}, \qquad \sigma = E(x)\varepsilon \tag{2.19}$$

The first equation is the equation of motion, the second is a consequence of the definitions of strain $\varepsilon = \partial u/\partial x$ and velocity $v = \partial u/\partial t$, while the last is Hooke's law. Both the elastic modulus E and the density ρ may vary with x, although for thermally induced variations along the Hopkinson pressure bars density variations are taken as negligible.

Elimination of ε from Eq. (2.19) gives a hyperbolic system of equations for stress and velocity:

$$\frac{\partial \sigma}{\partial x} = \rho(x)\frac{\partial v}{\partial t}, \qquad \frac{1}{E(x)}\frac{\partial \sigma}{\partial t} = \frac{\partial v}{\partial x} \qquad (2.20)$$

The characteristics of the equations are defined by

$$dx = c(x)\, dt \qquad \text{on which} \qquad d\sigma = \rho(x)c(x)\, dv \qquad (2.21)$$

$$dx = c - (x)\, dt \qquad \text{on which} \qquad d\sigma = -\rho(x)c(x)\, dv \qquad (2.22)$$

where

$$c(x) = [E(x)/\rho(x)]^{1/2} \qquad (2.23)$$

Numerical solutions to this hyperbolic set of equations can be used to calculate the incident and reflected pulses in the incident bar.

A closed-form solution for a temperature correction factor can be derived and is presented here following the analyses developed by Francis and Lindholm [32] for the case where the variation in density ρ is negligible. Although derived originally for an exponential temperature profile, the results apply to any temperature profile, as will be shown.

In Figure 2.31, if the characteristic HQ′ represents the characteristic giving the peak stress magnitude on $x = x_g$, say at Q′, the corresponding stress and velocity at the point C are related to quantities at F on the leading wave front by the integral of Eq. (2.22). Thus, since $v_F = 0$

$$v_C = -\int_F^C \frac{d\sigma}{\rho(x)c(x)} \qquad (2.24)$$

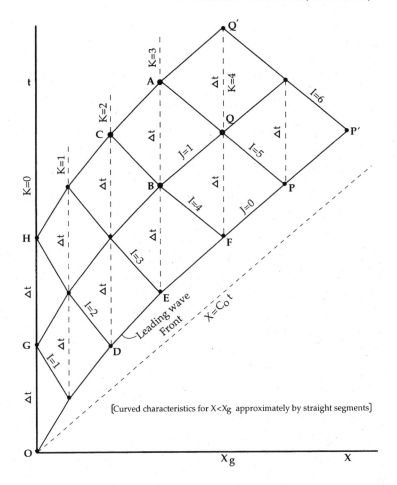

Figure 2.31 Characteristic mesh.

The integral in Eq. (2.24) is a line integral along the characteristic FC, which cannot be evaluated explicitly in general.

If now the pulse rise time is allowed to approach zero, approaching a shock wave of zero rise time, curve HQ' approaches OF, C approaches D, and $\rho(x)c(x)$ along the negative characteristic leading to C approaches $\rho(x_D)c(x_D)$ so that Eq. (2.24) approaches

$$v = -\sigma/\rho(x)c(x) \tag{2.25}$$

which relates v and ρ along the leading wave front just after passage of the discontinuous wave. Differentiation of Eq. (2.25) along the wave front gives

$$dv = \frac{\sigma}{(\rho c)^2} (\rho c)' \, dx - \frac{d\sigma}{\rho c}$$

where $(\rho c)'$ denotes $d(\rho c)/dx$. When this is substituted into $d\sigma = \rho c \, dv$ (Eq. (2.21)) to eliminate dv, the result is

$$\frac{d\sigma}{\sigma} = \frac{1}{2} \frac{(\rho c)'}{(\rho c)} \, dx \qquad (2.26)$$

This may be integrated explicitly along the wave front to obtain

$$\sigma/\sigma_i = (\rho c/\rho_i c_i)^{1/2} \qquad (2.27)$$

giving σ as a function of x just after the shock wave passage if ρc is a given function of x. The subscripts i refer to the interface end $x = 0$ of the bar. In particular, the stress σ_o at the gage stations where $\rho c = \rho_o c_o$ is given by

$$\sigma_o = \sigma_i (\rho_o c_o/\rho_i c_i)^{1/2}$$

or, since σ_o is measured and σ_i is to be calculated

$$\sigma_i = \sigma_o (\rho_i c_i/\rho_o c_o)^{1/2} \qquad (2.28)$$

Thus, we obtain the stress correction factor (SCF) as,

$$\text{stress correction factor} = (\rho_i c_i/\rho_o c_o)^{1/2} \qquad (2.29)$$

The strain correction factor is obtained by substituting Hooke's law

$$\sigma_i = E_i \varepsilon_i \qquad \text{and} \qquad \sigma_o = E_o \varepsilon_o$$

into Eq. (2.28). Thus,

$$\varepsilon_i = (E_o/E_i)(\rho_i c_i/\rho_o c_o)^{1/2} \varepsilon_o \qquad (2.30)$$

giving the

$$\text{strain correction factor} = (E_o/E_i)(\rho_i c_i/\rho_o c_o)^{1/2} \qquad (2.31)$$

When the density variation is negligible, substituting $c_i = (E_i/\rho_i)^{1/2}$ and $c_o = (E_o/\rho_o)^{1/2}$ into Eqs. (2.29) and (2.30) gives for $\rho_i = \rho_o$

$$\text{strain correction factor} = (E_o/E_i)^{-3/4}$$
$$\text{stress correction factor} = (E_i/E_o)^{1/4} \tag{2.32}$$

If the modulus is linearly dependent on the temperature as for example,

$$E_i = E_o(1 - C_\alpha) \tag{2.33}$$

where

$$C_\alpha = \alpha(T_i - T_o)$$

Then, the correction factors reduce to

$$\varepsilon_i/\varepsilon_o = (1 - C_\alpha)^{-3/4}$$

and

$$\sigma_i/\sigma_o = (1 - C_\alpha)^{1/4} \tag{2.34}$$

Application of these correction factors to the specimen strain and stress requires the further assumption that the correction factors, which were derived for the maximum strain and stress in a pulse of zero rise time, can be applied point by point during the finite rise time of a continuous pulse. If they can be so applied both in the incident bar and the transmitter bar, the integral of the reflected pulse, which gives the specimen strain by Eq. (2.12), will be corrected by the same strain correction factor as the pressure bar strains.

2.5 HOPKINSON PRESSURE BAR (TENSION)

- Materials
  Al-Al$_3$Ni, Ni-Nb-Al, FP/Al
  SiC/6061, SMC, CFR (polyester)
  KFR (polyester), glass/epoxy

Carbon/glass/epoxy
Carbon/Kevlar/epoxy
Woven carbon/epoxy
Graphite/epoxy
- Loading rates
  10^2–10^4/s
- Test configuration
  Tensile coupons
- Data obtained
  Strain sensitivity
  Stress pulse shaping
  Constituent properties
  (filament size, volume fraction)
  Dynamic yield/ultimate stress
  Fracture mechanisms
  Damage initiation

Considerable interest in the dynamic tensile testing of filamentary composites has occurred in recent years. While the equations for the incident, transmitted and reflected pulses are identical to those of the compression test, there are a number of important issues related to the overall design of the tensile pressure bar system, including specimen configuration and fixture. An early pressure bar apparatus capable of rates up to 2000/s was suggested by Harding et al. [33], and used for testing a number of metal-based materials. Such a configuration is shown schematically in Figure 2.32.

One of the earliest tests on composite type specimens tested in a tensile mode was performed by Matera and Albertini [20]. These tests were performed on eutectic type composites of Al–Al$_3$Ni and Ni–Nb–Al using a modified pressure bar apparatus shown schematically in Figure 2.33. Some typical strength versus strain rate data for Al–Al$_3$Ni composites is shown in Figure 2.34 for a range of temperatures. It has been noted that the response of this system to high strain rates is controlled mainly by the aluminum-rich matrix. Continued improvement in tensile pressure bar devices for high strain rate testing of metal-based materials has been suggested by Tatro et al. [34], and Nicholas [35,36]. Each of the systems shown schematically in Figures 2.35 and 2.36 represents a different configuration for intro-

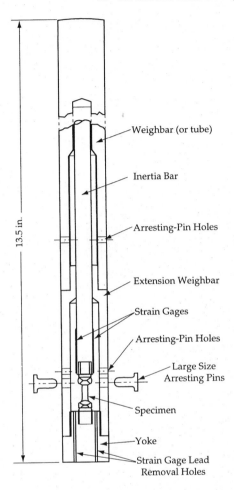

Figure 2.32 Tensile pressure bar apparatus [33]. (Reprinted with permission from the Council of the Institution of Mechanical Engineers, London.)

ducing both the dynamic stress into the specimen as well as the specimen configuration and fixturing, all important elements for filamentary composite specimens.

Results obtained from the first test procedure (Tatro et al. [34]) have shown that difficulties exist in the design of the specimen gripping as well as specimen size, leading to a scatter band of test data. A more reproducible tensile pressure bar apparatus for metals is that suggested by Nicholas [35] which uses a threaded type specimen. Data obtained for a wide variety of metals have been shown to be in good

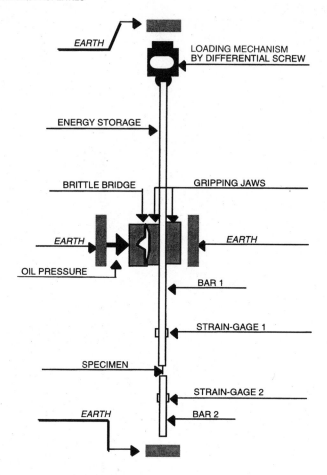

Figure 2.33 Schematic Hopkinson pressure bar system [20]. (Reprinted with permission from Elsevier Science S.A., Lausanne, Switzerland.)

agreement with other investigators. A modification of this tensile device for testing particulate, short fiber, and continuous fibre type composites was introduced by Ross et al. [37]. Particular attention was paid to the design of the specimen-holding configuration as shown schematically in Figure 2.37. For SiC/6061/Al, SMC, and FP/Al composites, which can be machined to produce threaded specimens, high strain-rate data were obtained. For some types of composites which cannot be machined, a smooth contoured waisted type specimen can be useful. Recently, Welsh and Harding [38] have conducted high-rate tensile tests using the pressure bar for a variety of

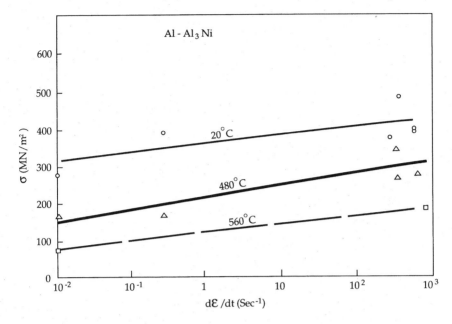

Figure 2.34 Strength vs. strain rate and temperature for Al-Al$_3$Ni [20]. (Reprinted with permission from Elsevier Science S.A., Lausanne, Switzerland.)

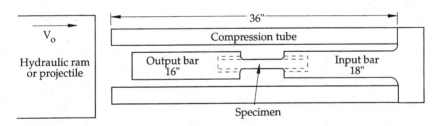

Figure 2.35 Schematic of tensile split Hopkinson pressure bar [34].

continuous filament composites. Materials tested include carbon, glass, and Kevlar-reinforced polyester resins. Some typical results obtained for several strain rates for carbon and glass are shown in Figure 2.38. It has been observed that there is an effect of strain rate on failure strength and the fracture appearance of composites for impact loading. An extension of tensile dynamic response for hybrid type composites was studied by Saka and Harding [39]. A typical waisted type specimen used in the studies is shown in Figure 2.39

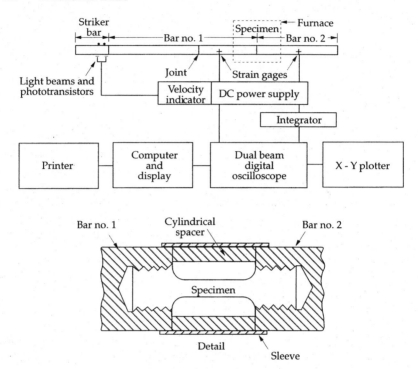

Figure 2.36 Schematic of tensile split Hopkinson pressure bar apparatus [35]. (Information reprinted courtesy of the Society for Experimental Mechanics, Inc.)

with a schematic of the SHPB arrangement and wave diagram shown in Figure 2.40. Some dynamic stress-strain curves obtained for various types of composites are shown in Figure 2.41. It has been observed by Saka and Harding [39] that fracture for carbon fiber-reinforced plastics show limited fiber pull-out with local cracking in the resin in planes perpendicular to the applied load. For glass fiber-reinforced plastics, an extended damage zone with no clearly defined fracture surface and considerable fiber pull-out has been noted. Additional studies on the behavior of glass/epoxy, graphite/epoxy and carbon/glass hybrids has been reported by Saka and Harding [40]. A schematic of the test configuration used in the study as well as typical results obtained for a glass weave composite relating stress to strain rate are shown in Figures 2.42 and 2.43. Of note in these results is that the yield stress shows a more significant increase with strain rate than does the maximum stress. Most recently, Liu and

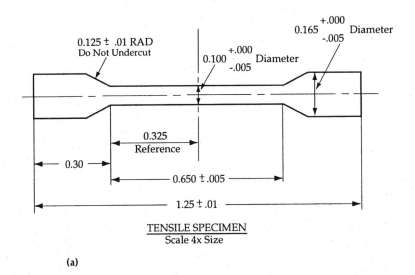

(a)

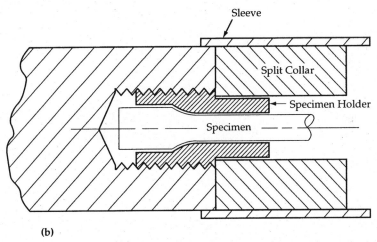

(b)

Figure 2.37 Specimen configurations Tensile SHPB FP/A1,Si.6060 A1, SMC composites [37]. (a) Machine drawing; (b) specimen inserted in bar. (Reprinted with permission from the Society for Experimental Mechanics, Inc.)

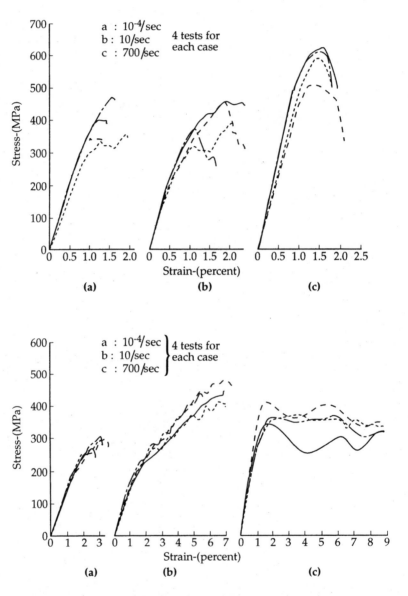

Figure 2.38   Tensile stress–strain curves for carbon fiber reinforced plastics and graphite fiber reinforced plastics [38]. (Reprinted with permission from the Institute of Physics Publishing Ltd.)

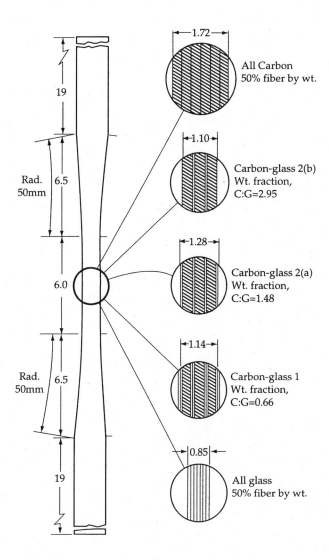

Figure 2.39 Test specimen configurations [39].

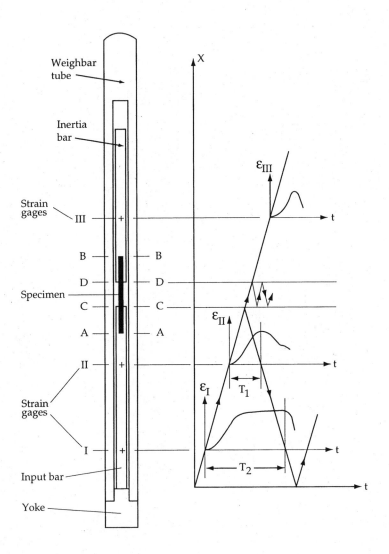

Figure 2.40 Schematic tensile SHPB arrangement and wave diagram [39].

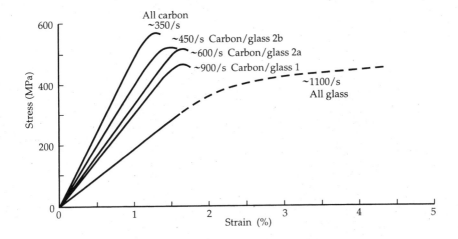

Figure 2.41 Dynamic stress–strain curves for various composites [39].

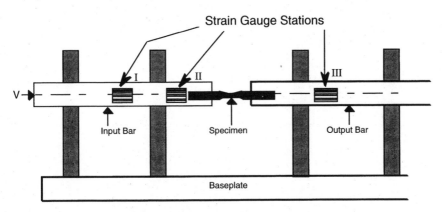

Test rig for impact compression tests

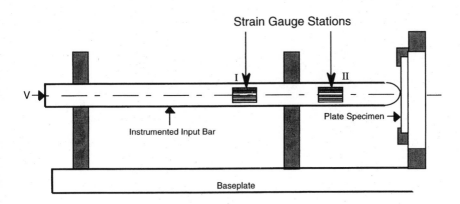

Figure 2.42 Test configuration pressure bar tests [40].

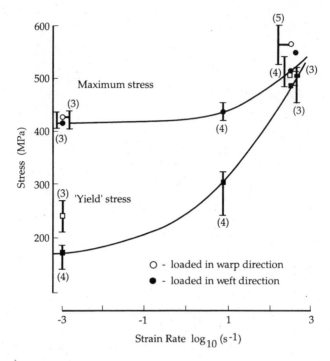

Figure 2.43 Stress vs. strain–rate for some composites [40].

Chiem [41] have proposed a tensile test device for composites patterned after the overall test device of Tatro et al. [34], and the specimen configuration/fixture of Ross et al. [37]. A schematic of the bar apparatus and Lagrangian wave diagram are shown in Figure 2.44 with typical stress-strain data obtained for woven carbon/epoxy composite shown in Figure 2.45.

2.6 HOPKINSON PRESSURE BAR (FLEXURE)

- Materials
  Steel
- Loading rates
  10^2–10^4/s
- Test configurations
  Beam specimen

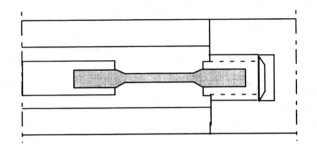

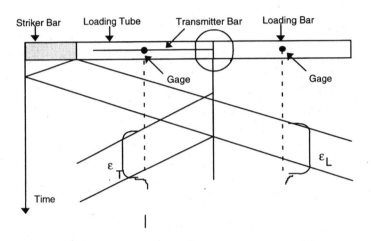

Figure 2.44 Schematic tensile SHPB system [41]. (Information reprinted courtesy of the Society for Experimental Mechanics, Inc.)

- Data obtained
  Fracture toughness
  Crack initiation
  Load time history

A modification of the Hopkinson pressure bar for evaluating the fracture toughness of materials has been studied by Mines and Ruiz [42] and Ruiz and Mines [8]. The experimental technique devised is shown schematically in Figure 2.46 with recorded strain–time histories also shown in Figure 2.46. As discussed previously, some difficulties that are associated with data reduction for Charpy tests are accountability for inertia forces within the specimen, output trace

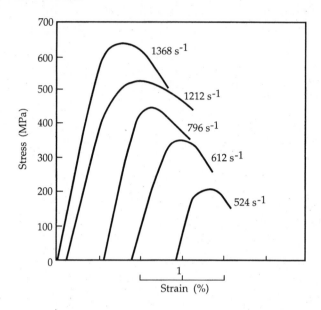

Figure 2.45 Tensile SHPB stress–strain curves woven carbon-epoxy composites [41]. (Information reprinted courtesy of the Society for Experimental Mechanics, Inc.)

oscillations, and the question as to whether specimen fracture occurs at the maximum applied load. Some of these difficulties can be avoided by using a pressure bar test technique. Specifically, the HPB provides a technique for obtaining a well-defined impulse load in which fracture occurs before reflection of the stress waves at the distal end of the bars. For this configuration, the load application is sufficient to encompass the time to fracture for the specimen and such occurrences as specimen movement in the loading anvils are eliminated.

The use of the pressure bar in this test mode follows the assumption of one-dimensional wave motion. As shown, the impact bar A in Figure 2.47 collides with the stationary bar B in contact with the test specimen. At point C a compression wave propagates. Strain gages located on bar B and separated by the distance L_{12} as denoted on bar B record the arrival of the compression wave. The wave speed in the bar is then given by

$$C = \sqrt{E/\rho} \qquad (2.35)$$

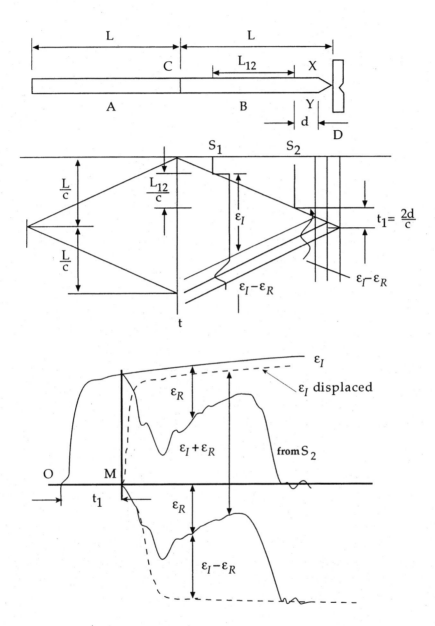

Figure 2.46 Flexure pressure bar apparatus and stress wave diagram [8]. (Reprinted with permission from Kluwer Academic Publishers.)

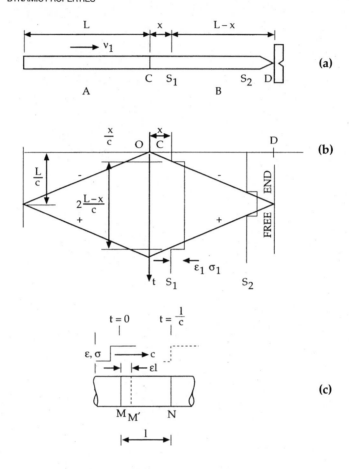

Figure 2.47 Schematic SHPB fracture apparatus, Lagrangian wave diagram, velocity and stress pulse relationship [8]. (Reprinted with permission from Kluwer Academic Publishers.)

where E is the modulus of elasticity of the bar and ρ the density of the bar. With bar B in contact with the specimen, the stress wave generated by the impacting bar A passes by gage stations S_1 and S_2, respectively, with an incident signal ε_I. The second strain gage S_2 unloads due to the reflected wave from the specimen. Thus, the gage S_2 sees the incident pulse ε_I and the reflected pulse ε_R. The time for the reflected wave to reach S_2 can be obtained using the equation

$$t = \frac{2d}{c} \tag{2.36}$$

where d represents the distance from the gage S_2 to the transition cone of the bar. Thus, at the time cited, gage station S_1 in the bar is still recording ε_I while gage S_2 is recording $(\varepsilon_I + \varepsilon_R)$. The reflected pulse can thus be obtained by subtracting out the two recorded strain-gage signals, giving $S_2 - S_1$. The force applied to the specimen can be obtained from the equation

$$P(t) = AE[\varepsilon_I - \varepsilon_R] \tag{2.37}$$

while the particle velocity in the bar is given by

$$v(t) = c[\varepsilon_I + \varepsilon_R] \tag{2.38}$$

By integrating the above equation the specimen deflection can be determined.

2.7   HOPKINSON PRESSURE BAR (SHEAR)

- Materials
  Polymethylene oxide
  Polycarbonate graphite/epoxy
  Polyamide
- Loading rates
  10^2–10^4/s
- Test configurations
  (short cylinder, thin-walled tubes, beams)
- Data obtained
  Strain sensitivity
  Stress pulse shaping
  Constituent properties
  (filament size, volume fraction)
  Dynamic yield/ultimate stress
  Fracture mechanisms
  Damage initiation

The pressure bar design discussed above represents a successful means of obtaining an assessment of the fracture toughness of materials as well as a means of obtaining the energy absorbed within the specimen at any given time.

One of the most important test modes for composites is the shear configuration. Torsional pressure bars have been developed for metals in order to eliminate friction and transverse stresses due to radial inertia effects present in axial loading. The launch mechanism used for torsional bars is generally either a torsionally loaded mechanical device or an explosive charge. The specimens used generally are thin-walled tubular specimens in lieu of the cylindrical disk specimens for compression testing or the dogbone type specimen for tensile testing. In general, for composites, the problem related to the size effects for compression and tensile loaded specimens is replaced by the difficulty of machining torsional specimens. For metals, Duffey [43] has used a torsional pressure bar to test thin-walled tubes.

A typical schematic of such a pressure bar configuration is shown in Figure 2.48. For the case shown, an explosive charge was used for initiating the pulse through the thin-walled tube. The two charges used to produce the torsional pulses were exploded simultaneously using equal weight charges with the resulting stress pulse so generated essentially torsional. Other pulses such as axial and bending are monitored, although these are considered to be negligible. Details of other techniques used for filtering the stress pulses generated can be found in Duffey [43].

To test polymer matrix materials, Vinh and Khalil [44] have devised a special torsional device using a motor-driven flywheel with an electromagnetic clutch as shown in Figure 2.49. This device ensures that the specimen strain rate is essentially constant since the flywheel-stored kinetic energy is greater than the energy absorbed in the specimen. The samples used as test specimens are short thin tubes and short cylinders. The data obtained have been used in the development of constitutive equations on the microstructural scale.

A torsional bar to measure large strains has been constructed by Stevenson [45]. In this apparatus, there is a mechanism which twists the specimen with a notched clamp bolt until fracture of the bolt occurs. Improvements in clamp design, bearing design and location, and specimen design have been introduced.

The dynamic testing of brittle materials, particularly those of the ceramic type, has been studied by Costin and Grady [46]. In these experiments a mechanical launch system using pre-torquing of the input bar has been used along with a solid cylindrical specimen. Analysis following that of the compressional pressure bar indicate

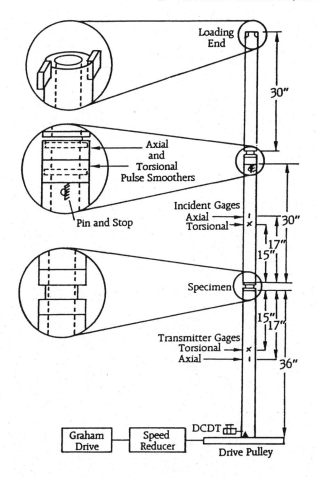

Figure 2.48   Torsional pressure bar apparatus [43].

that the torque applied to the specimen is equal to the measured torque, T_R, in the transmitter bar with the twist, θ, applied to the specimen proportional to the reflected pulse, that is,

$$\theta = \frac{2T_R(t)}{\rho Jc} \tag{2.39}$$

Here ρ is the pressure bar density, J the polar moment of inertia of the bars, and c the shear wave velocity.

A modification to the torsional pressure bar has been introduced by Gilat and Pao [47] using a mechanical launch mechanism. In this

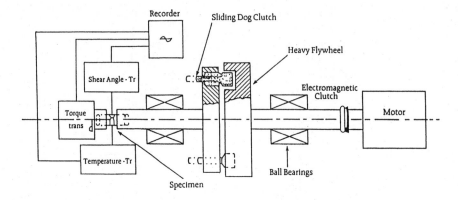

Figure 2.49 Dynamic torsion bar clutch apparatus [44]. (Reprinted with permission from the Institute of Physics Publishing Ltd.)

device, the input bar has been made of two sections of different cross-sectional areas. The largest cross-sectional bar area is located next to the loading wheel with a clamp placed on the smaller area input bar. A torque is applied at the clamp of the input bar and then the clamp is released. A torsional wave then propagates toward the specimen with a relief wave propagating towards the loading wheel. A typical Lagrangian wave diagram for the loading device described is shown in Figure 2.50.

Four strain gages have been positioned at the stations indicated in Figure 2.50 to monitor passage of the torsional wave. As with the case of compression and tension pressure bar design, if it is assumed that the specimen deformation is homogeneous, then the specimen strain rate can be described by

$$\dot{\gamma} = \frac{r}{L}\dot{\theta}_s \qquad (2.40)$$

In Eq. (2.40), r is the main radius of the thin-walled tube specimen, L is the specimen length, and $\dot{\theta}_s$ is the relative angular velocity between the two sides of the specimen. The relative angular velocity, $\dot{\theta}_s$, can be expressed in terms of the angular velocities at the specimen ends of the input and output bars as

$$\dot{\theta}_s = \dot{\theta}_i - \dot{\theta}_o \qquad (2.41)$$

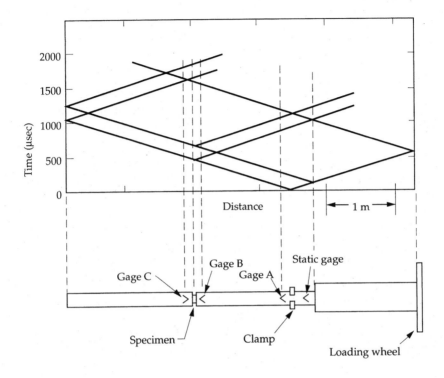

Figure 2.50 Torsional pressure bar and Lagrangian diagram [46]. (Information reprinted courtesy of the Society for Experimental Mechanics, Inc.)

The values of $\dot{\theta}_i$ and $\dot{\theta}_o$ can be measured in terms of the applied torques as

$$\dot{\theta}_i(t) = \frac{1}{\rho c J}\left[T_A\left(t - \frac{L_A}{c}\right) + T_A\left(t - \frac{L_A}{c} + \frac{2L_B}{c}\right) - T_B\left(t + \frac{L_B}{c}\right)\right]$$

(2.42)

$$\dot{\theta}_o(t) = \frac{1}{\rho c J} T_C\left(t + \frac{L_C}{c}\right)$$

(2.43)

where the quantities ρ, J, and c are the density, polar moment of inertia, and shear wave speed in the bar. The quantities L_A, L_B, and L_C are the distances between the strain gages and the specimen and T_A, T_B, and T_C are the torques recorded at gage stations A, B, and C, respectively. The shear stress at any time can be written as

$$\tau(t) = \frac{1}{2\pi r^2 h} T_C \left(t + \frac{L_C}{c} \right) \qquad (2.44)$$

where h is the wall thickness of the specimen.

The system described above is useful for testing specimens in the strain-rate range of 10^2–10^4 per second. Applications of this torsional pressure bar have been made in studies of composite cements as noted by Sierakowski et al. [48]. Some typical shear stress-strain curves for polymer hydraulic cement are shown in Figure 2.51.

It should be noted that while torsional pressure bars have been used to study the dynamic shear of metals, few tests have been directed to the study of composites. This is due in part to complication in the evaluation of laminated composites due to the importance of interlaminar and transverse shear effects. An application of the Hopkinson pressure bar for studying fiber-reinforced plastics in interlaminar and transverse shear modes has been suggested by Warner and Dharan [49]. In this experimental procedure, short beam specimens were introduced for two loading configurations. Details of the input and output pressure bars, as well as the specimen configuration, are shown in Figure 2.52. A concern with this test configuration, as well as with all pressure bar test modes, is the size of the composite specimens needed for obtaining meaningful test data. For the specimen sizes and test configurations studied by Warner and Dharan [49], it has been observed that the interlaminar shear remains relatively constant for all loading rates while the transverse shear decreases with increasing strain rate.

2.8   FLYER PLATE TESTS

- Materials
  Quartz-cloth-reinforced phenolic
- Loading rates
  10^4–10^6/s
- Test configurations
  Plates
- Data obtained
  Constitutive modeling parameters
  Pulse attenuation

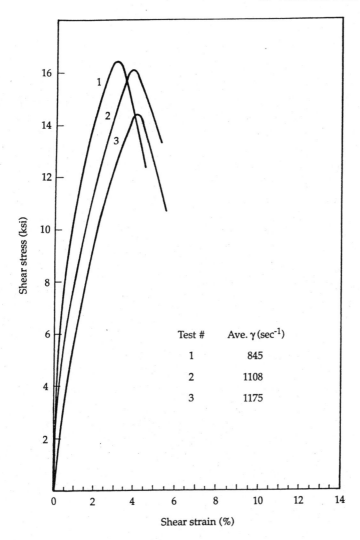

Figure 2.51 Dynamic shear stress–strain curves for hydraulic cement [48].

Dynamic fracture
Stress-wave-induced damage
Material properties degradation

Flyer plate tests have been introduced to study material behavior and failure modes at high stresses (>1 kbar loading) and short

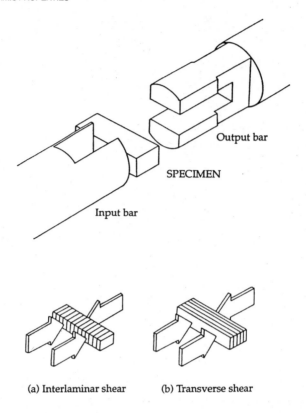

(a) Interlaminar shear   (b) Transverse shear

Figure 2.52 Shear pressure bar apparatus and specimen configurations [49]. (Reprinted with permission from Technomic Publishing Co., Inc.)

time-duration loads ($< 10^{-6}$ s). For such tests, the associated strain rates are 10^4 per second or higher and represent the highest strain rates attainable for a one-dimensional strain state. The technique used generally involves the impact of a flyer plate of one material projected at a target of a similar or different material. The most popular launch devices for such systems include gas guns, explosive discharge, exploding foil, and magnetic-driven propulsive devices. Each of these launch mechanisms can be considered as input in obtaining certain types of data. For example, the gas gun launch system is found useful for establishing constitutive modeling while the exploding foil technique produces higher velocities and is useful in establishing failure data. The basic technology associated with flyer plate testing has been reported by Berkowitz and Cohen [50], Duffey [43], and Rajendran and Bless [51].

Some of the issues that are important in the design of the flyer plate test system are,

- launch mechanisms;
- flyer support conditions;
- flyer velocity;
- flyer planarity;
- flyer dimensions and flight path;
- simultaneity of input;
- target support conditions;
- target dimensions and material characteristics;
- instrumentation of target;
- target recovery.

Some comments on each of these design issues follow.

A popular launch mechanism for throwing off flyer plates uses a confined electrical explosion to propel a thin metal foil producing high pressure. The foil is exploded by a capacitance discharge, and the velocity is controlled by changing the charge voltage. Figure 2.53 shows single-plate and double-plate flyer assemblies.

In the single-plate assembly, the foil explosion generates an essentially planar pressure pulse. The pressure pulse causes a plate type specimen to tear out of a material sheet (for example, Mylar) and be propelled towards the target. The free flight of the plate is usually of the order of 0.50 inches so that there is small tilt as the plate reaches the target. The stand-off area between the target and foil assembly is generally evacuated to a preset vacuum condition, for example 10^{-2} torrs, to minimize air cushion effects.

For the double flyer plate assembly, a single flyer plate coupled with a so-called moderator block is used. The first flyer plate is propelled into a moderator block to which a second flyer plate is attached by means of a lubricant. Upon impact of the first flyer plate, the second flyer plate is released with a free flight extending over approximately 0.25 in. Lucite is often used as the moderator material, and impact velocities appear to be well controlled with this material.

The target holder is generally made of a tool steel with the edges of the target wrapped with a layer of thin tape to prevent binding of the target to the holder. After impact, the targets are generally removed using a polystyrene foam or other soft material. During operation,

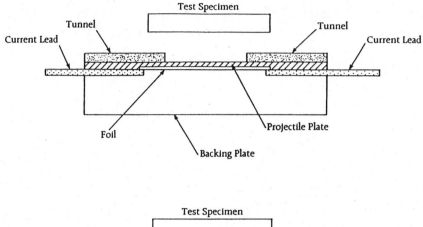

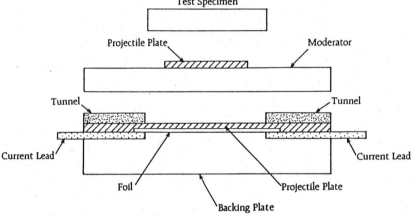

Figure 2.53 Single (top) and double flyer (bottom) plate assemblies.

care is exercised to ensure that the entire assembly is enclosed in a vented container.

The flyer plate velocity, prior to impact, is measured by change in closure of sequentially located shorting pins posted along the launch tube. Other techniques such as the cut-off of multiple-lighted beams have also been used.

Flyer plate planarity (tilt and distortion) is generally measured by the use of some type of pin-closure technique located around the periphery of the target. Alternatively measurement is made by photographing the flyer plate with a high-speed framing camera.

Typical flyer plate dimensions vary depending upon the launch techniques used. For gas propulsion, up to 4-in diameter specimens are used; for exploding foils, typical plate dimensions are 2 or 3 in. The use of magnetic propulsive techniques allow the accommodation of specimens which are curved, with Mylar and Lucite used as flyer plate materials.

Target support conditions are generally of the free or fixed variety. Free boundary conditions are simulated by placing the target material in a low-impedance holder; fixed boundary conditions are simulated by restraining the target material in a higher-impedance support holder. For composites, a fixed type of support condition is generally preferred.

Target dimensions conform to those of the flyer plates mentioned above. The specimens are normally sized so that edge effects can be ignored. That is, the lateral dimensions are selected to ensure that wave reflections from the plate boundary do not reach the central portion of the pressure pulse before observations are completed or damage to the target occurs. Target thickness is generally selected on the basis of the time required for a short propagating pulse with a velocity U_S to traverse the thickness distance L. By equating the respective transit times, and assuming a homogeneous isotropic material, we can obtain the following equation for sizing the flyer width to target thickness:

$$\frac{W}{L} = 2\{(\rho c/\rho_o U_s)^2 - 1\}^{1/2} \qquad (2.45)$$

Here, W is the flyer/target width, L the target thickness, ρ/ρ_o the material densities across the shock front, and U, the flyer velocity.

Target instrumentation generally consists of quartz transducers, manganin wires, capacitance gages, interferometers, and various high-speed framing and streak cameras. In composites, while some of the above techniques can be useful, they must be used with caution since material heterogeneity and anisotropy may involve more complicated states of strain, out of plane displacements, local and global failure events, and other factors.

Target recovery is one of the most difficult aspects of flyer plate testing. Some techniques allow the target to enter a container (tube) consisting of a soft material. Generally, the type of recovery system selected depends upon the type of observation to be made. Target recovery techniques are relatively unimportant when examining constitutive equation of state parameters or pulse attenuation data. On the other hand, when considering stress-wave-induced damage, particularly for composites, it must be ensured that secondary structurally induced damage does not occur.

The basic tenet for all such testing is that a unidirectional wave front is produced by the normal impact of a plane flyer upon a plane target. A schematic model of the input of one semi-infinite medium upon another is shown in Figure 2.54. For short times, the central regions of the flyer/target behave as a semi-infinite medium; for later times, the lateral boundary effects can produce complicated responses of the flyer and target plate structures. The actual time involved for this purpose is dependent upon such factors as the target lateral dimensions, flyer/target thicknesses, and properties of the respective materials.

For analysis, computer codes are often used for predicting wave propagation and failure. For composites, stress-wave-induced damage, produced by impact loading, generally falls into one of the following categories:

- delamination;
- material removal at the impacted face;
- material property degradation.

These events are graphically depicted in Figure 2.55, which shows the categories of stress-wave-induced damage.

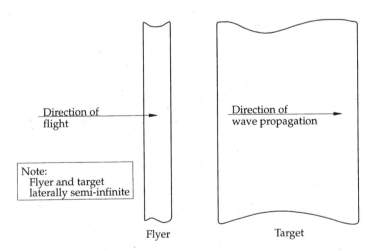

Figure 2.54 One-dimensional semi-infinite flyer plate experiment, schematic.

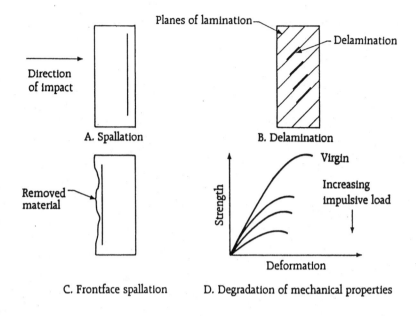

Figure 2.55   Stress-wave induced damage (schematic).

Stress-wave-induced damage is analytically predicted by considering a critical stress theory, a stress gradient theory, rate process criteria and empirical relationships. A summary of these criteria is included in Table 2.3 with specific comments on each of these criteria as follows.

The concept of a critical stress theory has been advanced which is tied to the breaking strength of a material, that is, a unique value of material strength denoted as a critical stress. Such information is obtained from the impact of a flyer plate into a target of the same material. If the material exhibits a constant critical stress, the impactor velocity will remain constant as the impactor/target thicknesses are increased or decreased in a fixed ratio.

Critical stress gradient theories are characterized by relations of the following types:

$$\sigma = A + B\left(\frac{\partial\sigma}{\partial x}\right)^{1/2} \tag{2.46}$$

where A and B are constants related to the material and pulse shape.

TABLE 2.3 Summary of dynamic fracture criteria [50]

Theory	Functional Relationship	Comments
Critical stress	$\sigma = \sigma_{cr}$	σ_{cr} = dynamic tensile strength
Stress gradient	$\sigma = A + B\left(\dfrac{\partial\sigma}{\partial x}\right)^{1/2}$	$A, B,\ \alpha, \beta$ = material and pulse shape constants
	$\sigma = \alpha + \beta\left(\dfrac{\partial\sigma}{\partial t}\right)^{1/2}$	σ = tensile stress
		x, t = spatial and time coordinates
	$(\sigma - \sigma_o)^\lambda \Delta t = K$	$K,\ \lambda$ = constants
	(1) or	σ = material strength constant
	$\displaystyle\int_o^t (\sigma - \sigma_o)^\lambda dt = K$	Δt = duration of tensile strength
Empirical relationships	(2) $I = \dfrac{k\sigma}{(\sigma - \sigma_o)^\lambda}$	I = tensile impulse
	(3) $\sigma - \alpha I^{-B}$	
		$\lambda,\ K,\ \alpha\ B$ = constants
Rate process criteria	$\sigma = A + B \log t$	
	$\dfrac{I}{I_1} = \dfrac{\sigma}{\sigma_1} e^{-\sigma/\sigma_1}$	$A,\ B,\ I,\ \sigma_1$ = material and pulse shape constants
	(1) or	
	$\sigma = \sigma_1 \ln(I_1/\sigma_1)$ $-\sigma_1 \ln(I/\sigma)$	

Empirical relationships can generally be approximated and written in the following forms:

$$I = K\sigma^{(1-\lambda)}$$

or

$$\sigma = \alpha I^{-B} \tag{2.47}$$

where K, λ, α and B are material constants which may not be related to physical processes.

Rate process theories are generally categorized by equations of the following type:

$$\sigma = c_1 + c_2 \log \dot{\varepsilon}$$

or

$$\sigma = c_3 + c_4 \log t \qquad (2.48)$$

With $\dot{\varepsilon}$ the strain rate, t the duration of the applied load, and the constants c_1, c_2, c_3, c_4 are dependent upon the material composition and temperature. The above equations are based upon experimental data and the constants cited above appear to be related in metals to the activation energy. Various dynamic fracture criteria proposed in the literature have been summarized in Table 2.3 [50].

2.9   CONCLUDING REMARKS

It should be noted that in the design of experiments for studying the dynamic behavior of composites, a clear distinction needs to be made between material response and structural response. This has been noted in a review by Harding [52] concerning the effect of high strain rates on material properties. In particular, material response is characterized by insensitivity to load application and specimen geometry while structural response is related to both specimen geometry and material properties. Thus, the design of experiments is compounded by the difficulty of distinguishing carefully between these two response modes.

As noted in the present chapter, dynamic test methods developed for obtaining high strain-rate data on the mechanical response of metals have been extended to composites. Since dynamic test data have been reported for metals, this can be considered important and relevant to understanding the dynamic performance of metal matrix composites. Rate-dependent data for non-metallic matrices have been studied, but there is less information on the rate dependency of the fiber phase of composites, due to the inherent difficulties in the testing of filaments and fiber bundles. The characterization of the constituents is important for assessing the performance of fiber-reinforced composites but the complex interaction between the reinforcing fibers and matrix phase leads to difficulties in assessing the rate dependency of the constituent phases. For example, when the rate of testing is increased, the corresponding failure mode changes. We conclude that some progress has been made in extending dynamic test techniques to composites and developing new rate-dependent tests for composites. However, assessing the mechanical behavior of composites will depend on identifying the response mode of the specimens. It is essential to identify the key geometrical and material properties features associated with the specimens. At issue, therefore, are the following factors: the processing variables involved in specimen fabrication; their quality control; the role of the fiber-matrix interface; environmental conditioning of the specimens; and selection of individual constituent phases, that is, the respective matrices and fibers selected for study.

The study of the effect of high strain rates on composite material properties is a rapidly developing field. Much of the development for composites is based upon understanding and experience gained from metals. For filamentary materials, an important class of composites, there are a number of important research issues: the complex interactions occurring between the fiber and matrix; the effect of fiber coatings on composite performance; the quantification of microdamage on composite response and failure. All of these issues reflect upon the difficulties associated not only with static strength measurements but also with the design of high strain-rate test devices. There are, of course, other important factors and parameters significantly affecting material strength, including time and temperature effects, moisture effects for polymer matrix composites, and the processing/fabrication aspects of various classes of composites.

2.10 EXPANDING RING

While the tensile Hopkinson pressure bar has been modified for use with composites, another high strain-rate technique has been developed based upon an expanding ring concept, as proposed by Hoggatt and Recht [53]. This technique has been modified for use with composite type specimens. While the technique offers an opportunity to obtain extremely high strain rates, it is difficult to determine the state of stress accurately. The technique does, however, provide for a uniform deformation in the specimen and avoids wave propagation effects. A schematic of this device as proposed by Hoggatt and Recht [53] for thin metal rings is shown in Figure 2.56.

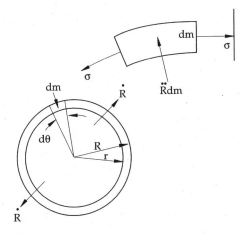

Figure 2.56   Dynamic symmetrical expansion of a thin ring [53]. (Information reprinted courtesy of the Society for Experimental Mechanics, Inc.)

The experimental concept involves the application of a short-time impulsive radial load to thin-ring specimens. This is accomplished by using a cylindrical steel ring; the expanding ring leaves a driver, and the ring continues to expand by virtue of its own inertia. An important element in the process is that symmetrical expansion of the ring occurs and that ring separation is achieved. The appropriate equations governing the process follow. For a thin-ring specimen as shown in Figure 2.56, the equation of motion in the radial direction for an element of the ring wall is given by [53]

$$-2\sigma(R-r)Z \sin \frac{d\theta}{2} = \rho(R^2 - r^2)Z_0 \ddot{R} \frac{d\theta}{2} \qquad (2.49)$$

where

$$\ddot{R} = \text{deceleration of the ring, in/s}^2$$
$$\sigma = \text{true stress, lbf/in}^2$$
$$\rho = \text{mass density, lb s}^2/\text{in}^4$$
$$R = \text{external radius, in}$$
$$r = \text{internal radius, in}$$
$$Z = \text{axial length of the ring, in}$$

Solving Eq. (2.49) for the hoop stress σ gives

$$\sigma = -\rho \frac{(R_o^2 - r_o^2)}{2(R - r)} \frac{Z_o}{Z} \ddot{R} \qquad (2.50)$$

By expressing r and Z as functions of R the hoop stress can be found from experimental measurements. Since unit strains in the axial and radial directions for an isotropic material will be equal,

$$\frac{Z_o}{Z} = \frac{R_o - r_o}{R - r} \qquad (2.51)$$

The volume of the ring is

$$V = \pi(R^2 - r^2)Z \qquad (2.52)$$

and for the uniaxial-stress conditions,

$$\frac{V}{V_o} = \left(1 + \frac{\sigma}{E}\right)\left(1 - \frac{\nu\sigma}{E}\right)^2 \qquad (2.53)$$

where

$$E = \text{Young's modulus in lbf/in}^2$$
$$\nu = \text{Poisson's ratio.}$$

Eqs. (2.51), (2.52), and (2.53) can be combined to give,

$$\frac{(R^2 - r^2)(R - r)}{(R_o^2 - r_o^2)(R_o - r_o)} = \left(1 + \frac{\sigma}{E}\right)\left(1 - \frac{\nu\sigma}{E}\right)^2 \qquad (2.54)$$

and the above cubic equation, when solved for the internal radius r, gives

$$r = \frac{R}{3}[1 + 4\cos(\phi/3 + 240°)] \qquad (2.55)$$

where

$$\cos \phi = -1 + \frac{26(R_o^2 - r_o^2)(R_o - r_o)(1 + \sigma/E)(1 - \nu\sigma/E)^2}{16R^3} \quad (2.56)$$

and $\phi =$ an angle lying in the second quadrant (that is, $\phi \leq 180°$).

Replacing Z_o/Z by Eq. (2.51) and r by Eq. (2.55) in Eq. (2.50) leads to the following equation for the true hoop stress in terms of the external radius:

$$\sigma = -\rho R \ddot{R} \left[\frac{[1 - (r_o/R_o)^2][1 - r_o/R_o]}{\frac{8}{9}(R/R_o)^3[1 - 2\cos(\phi/3 + 240°)]^2} \right] \quad (2.57)$$

where, for further discussion, the bracketed term is identified as $F(R, \phi)$.

Note that the independent variable ϕ appears on the right hand side of this equation. However, it can be shown that the value of ϕ is insensitive to the value of σ, and nominal values for stress can be used to determine the magnitude of $F(R, \phi)$ in the above equation. For thin rings, the magnitude of this term is nearly unity, although the term is sensitive to r_o/R_o and is insensitive to differences in material properties and to strain as represented by R/R_o [53]. For thin rings, the true hoop stress can be approximated by

$$\sigma = -\rho R \ddot{R} \quad (2.58)$$

In cases where experimental resolution is sufficient to justify a more accurate value, the function $F(R, \phi)$ can be applied as a correction factor to this equation in line with Eq. (2.57). As an example, a typical value of r_o/R_o is 0.97, which corresponds to a correction factor of 0.98, a 2% reduction in the stress as computed by Eq. (2.58).

The true (logarithmic) strain is given by

$$\varepsilon = \ln \frac{R}{R_o} \quad (2.59)$$

from which the true strain rate is given by

$$\dot{\varepsilon} = \frac{\dot{R}}{R} \quad (2.60)$$

Data in support of the above analysis are obtained by use of a high-speed streak camera. Attention to experimental detail that is needed for data acquisition are symmetry in expansion of the ring and accurate measurement of the displacement as a function of time.

An extension of this principle to graphite/epoxy composites with different ply constructions has been studied by Daniel et al. [54]. A schematic of the device is shown in Figure 2.57. In this test technique an explosively generated pressure pulse has been applied through a liquid medium to the specimen. Since the ring is thin, the wave takes only a short time to travel through the ring thickness. The thin-ring specimens used are generally made by cutting thin rings from cylindrical specimens. Dynamic pressures are measured on the side of the cylinder using a piezoelectric transducer and recorded on a digital processing oscilloscope.

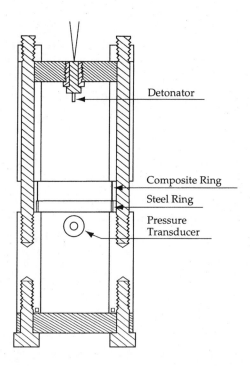

Figure 2.57 Schematic fixture for dynamic loading of composite-ring specimen [54]. (Information reprinted courtesy of the Society for Experimental Mechanics, Inc.)

In order to analyze the mechanics of a dynamically pressurized ring, the equation of motion for axisymmetric loading is used

$$\frac{\partial \sigma_r}{\partial r} + \frac{\sigma_r - \sigma_\theta}{r} = \rho \ddot{u} \qquad (2.61)$$

where σ_r, σ_θ are the radial and circumferential stress, r the radial distance, ρ the material mass density, and u the radial displacement. Using the Lamé equations for an internally pressurized cylinder, we find

$$\sigma_r = \frac{pa^2}{b^2 - a^2}\left(1 - \frac{b^2}{r^2}\right)$$

$$\sigma_\theta = \frac{pa^2}{b^2 - a^2}\left(1 + \frac{b^2}{r^2}\right) \qquad (2.62)$$

The equation of motion at $r = b$ can be written as

$$P\left(\frac{2a^2}{b^2 - a^2}\right) = \sigma_\theta + \rho b \ddot{u} = \sigma_\theta + \rho b^2 \ddot{\varepsilon}_\theta \qquad (2.63)$$

Here, a and b are the inner and outer radii of the ring. The primary data recorded in the dynamic ring tests are:

ε_θ^s = circumferential strain in the steel ring (pressure cell) at the outer radius, b;

ε_θ^c = circumferential strain in the composite ring at the outer radius, b;

ε_x^c = axial strain in the composite ring at the outer radius, b.

Assuming a uniaxial state of stress in the circumferential direction, the stress in the elastic ring can be computed from

$$\sigma_\theta^s = E^s \varepsilon_\theta^s \qquad (2.64)$$

The dynamic pressure is obtained in terms of this stress and the second derivative of ε_θ^s as

$$p = \left(\sigma_\theta^s + \rho_s b_s^2 \ddot{\varepsilon}_\theta^s\right)\left(\frac{b_s^2 - a_s^2}{2a_s^2}\right) \tag{2.65}$$

where E^s and ρ_s are the modulus and density of the steel, respectively.

The circumferential stress in the composite ring is then obtained in terms of the pressure and the second derivative of ε_θ^c as

$$\sigma_\theta^c = p\left(\frac{2a_c^2}{b_c^2 - a_c^2}\right) - \rho_c b_c^2 \ddot{\varepsilon}_\theta^c \tag{2.66}$$

The dynamic stress-strain curve for the composite material is obtained by plotting σ_θ^c vs ε_θ^c. Thus, the moduli, Poisson's ratios, strength, and ultimate strain for the composite material can be obtained from the recorded and computed data.

For the case of 10° off-axis rings used for determination of in-plane shear properties, three-gage rosettes have been mounted on the outer surface of the composite rings. The measurements taken were:

$$\varepsilon_\theta^s = \text{circumferential strain in steel ring}$$

$$\varepsilon_\theta^c = \text{circumferential strain in composite ring}$$

$$\varepsilon_x^c = \text{axial strain in composite ring}$$

$$\varepsilon_{45}^c = 45° \text{ strain in composite ring}$$

The circumferential stress in the composite ring was obtained from Eq. (2.66) with the in-plane shear stress, referred to the fiber direction, given by

$$\sigma_{12}^c = \sigma_\theta^c \sin \phi \cos \phi \tag{2.67}$$

where $\phi = 10°$, the fiber orientation with respect to the circumferential direction. The in-plane shear strain is obtained from the three strain components as follows:

$$\varepsilon_{12}^c = \frac{\varepsilon_\theta^c - \varepsilon_x^c}{2} \sin 2\phi + \left[\varepsilon_{45}^c - \frac{\varepsilon_x^c + \varepsilon_\theta^c}{2}\right] \cos 2\phi \tag{2.68}$$

A dynamic shear-stress vs shear-strain curve is obtained by plotting σ_{12}^c vs ε_{12}^c. The shear modulus is then given by

$$G_{12} = \frac{\sigma_{12}}{2\varepsilon_{12}} \tag{2.69}$$

and can be obtained from the initial slope or the secant of the stress-strain curve.

A computer program has been used which smoothes and approximates the strain, strain rate, and acceleration data. In this way, dynamic stress-strain curves can be obtained.

2.11 REFERENCES

1. Shieh, R.C. (1978) "Empirical equations for puncture analysis of lead-shielded spent full shipping casks," *Proceedings of Fifth International Symposium on Packaging and Transportation of Radioactive Materials*, p. 341.

2. Harding, J. (1979) "The high speed punching of woven-roving glass reinforced composites," *Mechanical Properties at High Rates of Strain*, Institutes of Physics Conference, Oxford, pp. 318–330.

3. Duffey, T.A., Glass, R.E., and Sutherland, S.H. (1984) "Response of woven Kevlar panels to low speed impact loading," *Institute of Physics Conference*, Oxford, pp. 549–550.

4. Marin, J. (1962) *Mechanical Behavior of Engineering Materials*, Prentice-Hall, Inc., Englewood, NJ.

5. Chamis, C.C. (1976) "Failure mechanics of fiber composite notched Charpy specimens," *Proceedings of the Army Symposium on Solid Mechanics, 1976—Composite Materials: The Influence of Mechanics of Failure on Design*, AMMRC, Watertown, MA.

6. Beaumont, P.W.R., Riewald, P.G., and Zweben, C. (1975) "Methods for improving the impact resistance of composite materials," *Symposium, Foreign Object Impact Damage to Composites*, ASTM STP 568, pp. 134–158.

7. Krinke, D.C., Barber, J.P., and Nicholas, T. (1978) "The Charpy impact test as a method for evaluating impact resistance of composite materials," AFML TR 78-54, Dayton, OH.

8. Ruiz, C. and Mines, R.A.W. (1985) "The Hopkinson pressure bar: an alternative to the instrumented pendulum for Charpy tests," *Int. J. Fracture*, **29**, 101–109.

9. Broutman, L.J. and Rotem, A. (1975) "Impact strength and toughness of fiber composite materials," *Foreign Object Damage to Composites*, ASTM STP 568, pp. 114–133.

10. Lifshitz, J.M. (1976) "Impact strength of angle ply fiber reinforced materials," *J. Composite Materials*, **10**, 92–101.

11. Caprino, G. "Residual strength prediction of impacted CFRP laminate," *J. Composite Materials*, **18**, 508–518.

12. Marom, G., Drukker, E., Weinberg, A., and Benloji, J. (1986) "Impact behavior of carbon/Kevlar composites," *Composites*, **17**, 150–153.

13. Adams, D.F. (1977) "Impact response of polymer-matrix composite materials," *Composite Materials: Testing and Design [4th conference]*, American Society for Testing and Materials, ASTM STP 617, pp. 45–58.

14. Ireland, D.R. (1981) "Instrumental impact testing for evaluating end-use performance," *Physical Testing of Plastics-Correlation with End-Use Performance*, American Society for Testing and Materials, ASTM STP 736, pp. 45–58.

15. Winkel, J.D. and Adams, D.F. (1985) "Instrumental drop weight impact testing of cross-ply and fabric composites," *Composites*, **16**(4), 268–278.

16. Kressler, S.L., Adams, G.C., Driscoll, S.B., and Ireland, D.R. (1986) "Instrument testing of plastics and composite materials," American Society for Testing and Materials, ASTM STP 936, Philadelphia, PA.

17. Kakarala, S.N. and Roche, J.L. (1986) "Experimental comparison of several impact test methods," American Society for Testing and Materials, ASTM STP 936, Philadelphia, PA.

18. Chou, S.C., Robertson, K.D., and Rainey, J.H. (1973) "The effect of strain rate and heat developed during deformation on the stress–strain curve of plastics," *Experimental Mechanics*, **13**, pp. 422–432.

19. Armenakas, A. and Sciammarella, C.A. (1973) "Response of glass-fiber reinforced epoxy specimens to high rates of tensile loading," *Experimental Mechanics*, **13**, 433–440.

20. Matera, R. and Albertini, C. (1978) "The mechanical behavior of aligned eutectics at high rates of strain," *Mater. Sci. Engng.*, **32**, 267–276.

21. Hopkinson, B. (1921) *Collected Scientific Papers*, Cambridge University Press, Cambridge.

22. Kolsky, H. (1953) *Stress Waves in Solids*, Clarendon Press, Oxford, p. 211.

23. Davies, E.D.H. and Hunter, S.C. (1963) "The dynamic compression testing of solids by the method of the split Hopkinson pressure bar," *J. Mech. Physics of Solids*, **11**, 155–179.

24. Lindholm, U.S. and Yeakley, L.M. (1964) "Dynamic deformation of single and polycrystalline aluminum," *J. Mech. Physics of Solids*, **13**, 41–53.

25. Larsen, T.L., Rajnak, S.L., Hauser, F.E., and Dorn, J.E. (1964) "Plastic stress/strain-rate/temperature relations in H.C.P. Ag–Al under impact loading," *J. Mech. Physics of Solids*, **12**, 361–376.

26. Conn, A.F. (1964) "On the use of thin wafers to study dynamic properties of metals," *J. Mech. Physics of Solids*, **12**, 361–376.

27. Hauser, F.E. (1966) "Techniques for measuring stress–strain relations at high strain rates," *Experimental Mechanics*, **6**, 395–402.

28. Bertholf, L.D. and Karnes, C.H. (1975) "Two-dimensional analysis of the split Hopkinson pressure bar system," *J. Mech. Physics of Solids*, **23**, 1–19.

29. Chin, S.S. and Neubert, V.H. (1967) "Difference method for wave analysis of split Hopkinson pressure bar with a viscoelastic specimen," *J. Mech. Physics of Solids*, **14**, 117–193.

30. Bell, J.F. (1966) "An experimental diffraction grating study of the quasi-static hypothesis of the split Hopkinson bar experiment," *J. Mech. Physics of Solids*, **14**, 309–327.

31. Perzyna, P. (1959) "The problem of propagation of elastic–plastic waves in a non-homogeneous bar," in *Non-Homogeneity in Elasticity and Plasticity*, Editor, W. Olszak, Pergamon Press, London New York, pp. 431–438.

32. Francis, P.H. and Lindholm, U.S. (1968) "Effect of temperature gradients on the propagation of elastoplastic waves," *J. Appl. Mech.*, **35**, 441–448.

33. Harding, J., Wood, E.O., and Campbell, J.D. (1960) "Tensile testing of materials at impact rates of strain," *J. Mech. Engng. Sci.*, **2**, 88–96.

34. Tatro, C.A., Gust, W.H., and Taylor, A.R. (1980) *High Rate Testing at Elevated Temperature*, UCRL-18637, Lawrence Livermore Laboratory.

35. Nicholas, T. (1981a) "Tensile testing of materials at high rates of strain," *Experimental Mechanics*, **21**, 177–185.

36. Nicholas, T. (1981b) "Material behavior at high strain rates," *Impact Dynamics*, John Wiley, New York, Chap. 8.

37. Ross, C.A., Cook, W.H., and Wilson, L.L. (1984) "Dynamic tensile tests of composite materials using a split-Hopkinson pressure bar," *Experimental Techniques*, **8**, 30–33.

38. Welsh, L.M. and Harding, J. (1984) "Effect of strain rate in the tensile failure of woven reinforced polyester resin composites," *Institute of Physics Conference*, Oxford, England, pp. 343–344.

39. Saka, K. and Harding, J. (1985) *Behaviour of Fibre-Reinforced Composites Under Dynamic Tension*, Report No. OUEL 1602/85, Oxford University.

40. Saka, K. and Harding, J. (1986) *Behaviour of Fibre-Reinforced Composites Under Dynamic Tension*, AFOSR-85-0218.
41. Liu, Z.G. and Chiem, C.Y. (1988) "A new technique for tensile testing of composite materials at high strain rates," *Experimental Techniques*, **12**, 20–21.
42. Mines, R.A. and Ruiz, C. "The dynamic behavior of the instrumented Charpy test," *J. de Physique*, **46**, 187–196.
43. Duffey, T.A. (1982) *The Dynamic Plastic Deformation of Metals—A Review*, AFWAL-TR-82-4024.
44. Vinh, T. and Khalil, T. (1984) "Adiabatic and viscoplastic properties of some polymers at high strain and high strain rate," *Institute of Physics Conference*, Oxford, England, pp. 39–46.
45. Stevenson, M.G. (1984) "Further development and use of a torsional Hopkinson-bar system for stress–strain measurements to large strains," *Institute of Physics Conference*, Oxford, England, pp. 167–174.
46. Costin, L.S. and Grady, D.E. (1984) "Dynamic fragmentation of brittle materials using the torsional Kolsky bar," *Institute of Physics Conference*, Oxford, England, pp. 321–328.
47. Gilat, A. and Pao, Y.H. (1988) "High rate incremental strain rate test," *Experimental Mechanics*, **28**(3), 322–325.
48. Sierakowski, R.L., Gilat, A., and Wolfe, W.E. (1987) "High strain rate of testing hydraulic cements," *Proceedings Engineering Mechanics Specialty Conference*, State University of New York, Buffalo.
49. Warner, S.M. and Dharan, C.K.H. (1986) "The dynamic response of graphite fiber-epoxy laminates at high shear strain rates," *J. Composite Materials*, **20**, 365–374.
50. Berkowitz, H.M. and Cohen, L.J. (1969) *A Study of Plate-Slap Technology*, AFML-TR-69-106.
51. Rajendran, A.M. and Bless, S.J. (1985) *High Strain Rate Material Behavior*, AFWAL-TR-85-4009.
52. Harding, J. (1986) *The Effect of High Strain Rate on Material Properties*, Oxford University Engineering Laboratory Report OUEL 1627.
53. Hoggatt, C.R. and Recht, R.F. (1969) "Stress–strain data obtained at high rates using an expanding ring," *Experimental Mechanics*, **9**, 441–448.
54. Daniel, I.M., Labedz, R.H., and Liber, T. (1981) "New method for testing composites at very high strain rates," *Experimental Mechanics*, **21**, 71–77.

CHAPTER 3

WAVE MOTION

3.1 INTRODUCTION

Fiber-reinforced composites used for structural components can be analyzed and designed at different material/structural levels based upon scale factors related to the constituent elements. For example, a unidirectional reinforced lamina is often considered as a homogeneous anisotropic continuum, disregarding constituent phases and their inherent heterogeneity. This idealization has proved to be extremely useful in situations involving quasi-static as well as some dynamic loadings. As another example, the thermoplastic properties of a composite system evaluated experimentally using a "lamina" as the basic specimen configuration are, at best, considered to be the global properties of the composite system. Recognizing that structural components used in real-life applications are made of laminates, consisting of a tailored stacking of laminae, lamina can be considered as the basic "building blocks" for analyzing the static as well as the dynamic response of composite laminates. It can be shown that this idealization has some serious drawbacks in dynamic events and particularly in situations where the microstructure (that is, the fibers and matrix phases with their widely different properties) begins to influence significantly the wave motion within a lamina. The phenomena of wave dispersion, scattering, and filtering of certain frequencies cannot be explained by idealizing the lamina to be homogeneous, that is without a microstructure but as an anisotropic medium. In the case of laminates, the interface existing between two adjacent laminae creates an additional microstructure which needs to be accounted for before a complete understanding of wave motion within fiber-reinforced composite materials can be obtained.

A vast list of reference sources is available and can be used to obtain information on various aspects of wave motion in a wide variety of composite materials. These references consider both steady-state and transient wave motions using theoretical models and experimental investigations. The various theories which have been proposed range from simple approximate models to very sophisticated ones, yet an exact understanding of the various dynamic aspects of composite behavior, especially under impact type of loadings, is still forthcoming. The present chapter is not intended to present in full, or even in part, the voluminous literature published on the subject. Some excellent survey articles, such as references [1–6], can be used as baseline sources of information. The discussion in this chapter is largely drawn from these articles and the papers referred to therein. Generally, the various models and theories as proposed in the literature can be categorized under a number of theoretical models. These models are essentially approximate and can be used to provide reasonably good results for one or more aspects of wave phenomena in heterogeneous composites. Some specific remarks and comments are noted for each of these theories with recognition that a detailed description of even one representative theory would merit a separate monograph on the subject. However, details of models which are simple yet accurate enough to deal with low-velocity impact loading situations are discussed as well as some relevant experimental investigations.

3.2 PRIMARY THEORETICAL MODELS

The primary models as discussed in references [1–6] are as follows:

- effective modulus theories;
- effective stiffness theories;

- elasticity theories with microstructure;
- mixture theories;
- continuous theories based upon asymptotic expansion;
- discrete continuum theories;
- lattice-type theories;
- other theories.

3.2.1 Effective Modulus Theories

These theories are based upon a primary assumption that as in the case similar to that of static response, the wave motion in a composite system can be analyzed by idealizing heterogeneous fiber reinforced composites as homogeneous anisotropic continua. In this idealization, the matrix and fiber phases (the constituents) are disregarded, and the effective elastic properties of the continuum are replaced by a geometrically weighted average of the properties of the constituents. Theories in this category include those of Postma [7], White and Angona [8], Weitsman [9], Behrens [10], Aboudi and Weitsman [11], and Sun et al. [12]. These theories exhibit some serious shortcomings in that the experimentally observed dispersion and attenuation of waves cannot be analyzed. These phenomena become important when the dominant signal wave lengths are of the order of the fiber diameters or the fiber-to-fiber spacing.

3.2.2 Effective Stiffness Theories

The effective stiffness theory is a continuum theory of composites that reflects the influence of fiber and matrix phases. In a typical theory [13], the fiber and matrix phases are assumed to be linearly elastic, isotropic and homogeneous, and the influence of these phases is accounted for through the assumed approximate displacements in each of the constituents. The interactions between the constituents can then be obtained by imposing displacement continuity conditions at the interface between the fibers and matrix. This formulation involves some smoothing operations involving terms in the strain energy density and kinetic energy functions with the application of Hamilton's principle for obtaining the best approximate solution. The theory appears to predict reasonably well the phenomenon of geometric dispersion over a significant range of wave numbers. Researchers using this approach include Sun et al. [13], Achenbach et al. [14], Grot and Achenbach [15-16], and Grot [17]; higher-order refinements for a laminated medium have been studied by Drumheller and Sutherland [18]; and a nonlinear theory introduced by Rausch [19].

3.2.3 Elasticity Theory with Microstructure

Many continuum theories dealing with fiber-reinforced composites implicitly assume the existence of a periodic microstructure in the form of equally spaced fibers. In a continuum lacking such periodicity, Mindlin [20] proposed a theory known as an elasticity theory with microstructure provided the matrix behaves as an isotropic elastic material. This theory was extended by Ben-Amoz [21] to account for inclusions of arbitrary geometry in the matrix material. Both theories consider a representative volume element (RVE) or a unit cell as a building block, with material constants including phase moduli and volume fractions. Thus, phenomena like wave dispersion in composites can be studied.

3.2.4 Mixture Theories

Mixture theories used to study wave motions in composites are based upon the fundamental work of Truesdell and Toupin [22], Green and Naghdi [23], Steel [24], and others. In a medium which is a mixture of two or more distinct phases (components), these components are allowed to undergo individual deformations. The constitutive formulation for the mixture requires superposition of components in space, however, specification of appropriate interaction parameters between the constituents on the basis of the knowledge of their geometrical and material properties poses serious difficulties. Models proposed by Bedford and Stern [25–27], Hegemier and Nayfeh [28], and Hegemier et al. [29] appear to be successful in incorporating these effects. These models appear to be suitable for studying both the harmonic and transient waves in fiber-reinforced or bilaminate type composites. The extension of these theories to include non-linear effects in composites is also feasible.

3.2.5 Continuum Theories Based upon Asymptotic Expansion

The models proposed by Ben-Amoz [30], Hegemier and Nayfeh [28], Hegemier et al. [29], Hegemier and Bache [31,32] and others [1] appear to provide quite reasonable results for laminated composites including both wave-guide and wave-reflective types of problems. These are micromechanical theories in the sense that they provide information on the stress and displacement distributions within each of the constituents. These approaches are found useful for evaluating steady-state as well as transient vibratory response, and can be extended to study both non-linear and non-periodic composites.

3.2.6 Discrete Continuum Theories

The discrete continuum theory, as proposed by Chao and Lee [33], defines the displacement field in the form of a two-term truncated Taylor series for each layer of a periodically layered composite. Continuity conditions at the interface between two adjacent layers are enforced in order to obtain governing equations in the form of differential-difference equations. Numerical results obtained for plane harmonic waves in an unbounded layered media have been found to agree quite closely with exact results, and even better agreements have been found as the wave number is increased when compared with results obtained using the effective stiffness theory [1].

3.2.7 Lattice-Type Theories

Due to the uniform nature of fiber orientation in fiber-reinforced composites, the waves moving in the fiber direction are channeled by the fibers acting as wave guides, while these fibers create an obstruction for waves propagating in their normal direction. This situation is analogous to the behavior of mass particles in a lattice system. A three-dimensional theory proposed by Turhan [34] models the fibers as long and slender structural elements with the surrounding matrix considered as a system of springs with a given stiffness. Drumheller and Sutherland [18] have applied lattice models to the case of laminated plates.

3.2.8 Other Theories

For waves propagating in periodic composites, variational methods have been used as analytical tests by Nemat-Nasser [35]. This approach has been used to obtain the basic field equations from consideration of the first variation of appropriate integrals and relies heavily on the appropriate selection of test functions for stresses and displacements. This can lead to serious complications when extended to transient problems and/or non-periodic composites.

A method proposed by Barker [36], based upon a viscoelastic analogy where the parameters involved are defined in terms of geometrical variables and constituent properties has been used for layered composites.

In order to deal with compressive shock loading in composites, theories based upon hydrodynamic concepts developed by Tsou and Chou [37,38], Torvick [39], and Munson and Schuler [40] are available. Other approaches that deal with wave motions in basically heterogeneous composites using statistical models have been reported by McCoy [41], Bose and Mal [42], Ziegler [43], and Krumhansl [44].

With the brief description outlined above for the various concepts and models that have been advanced to deal with the complex phenomena associated with wave motions in heterogeneous and anisotropic composites, some procedures and key results based primarily upon the effective modulus theories for continuous fiber reinforced polymer composites are discussed.

3.3   EFFECTIVE MODULUS THEORIES FOR COMPOSITES

The system selected for study is a unidirectionally reinforced composite lamina. This lamina is essentially non-homogeneous and anisotropic in its elastic properties, but for the purpose of analysis can be idealized as a homogeneous and anisotropic medium. In this idealization, the microstructure present in the composite, in the form of fiber and matrix phases with their different elastic properties, is disregarded, and the effective properties of the composite are represented as the geometrical weighted properties of the constituents, suitably chosen through representative volume elements within the composite continuum. This effective modulus idealization appears to be an adequate representation (model) for situations in which the effective wavelengths of the waves propagating through the composite are of the order of a hundred fiber diameters. With this idealization in mind, a number of the well established theories of wave propagation through an anisotropic solid continuum can be applied [45–47].

In this section, the general theory of plane elastic waves propagating through a homogeneous and anisotropic continuum is discussed with particular emphasis focused on an orthotropic elastic medium such as a continuous fiber-reinforced unidirectional composite lamina.

Recall the Cauchy equations of motion with zero body forces given by [47]

$$\sigma_{ij,j} = \rho \ddot{u}_i \tag{3.1}$$

Here, σ_{ij} refers to the stress components with respect to a right-handed Cartesian coordinate system x_i while the $\ddot{u}_i$ are the acceleration components, and ρ is the mass density of the body.

Considering small deformation theory, the strain-displacement relations can be written as

$$\varepsilon_{kl} = \tfrac{1}{2}(u_{k,l} + u_{l,k}) \tag{3.2}$$

It should be noted that throughout this chapter standard index notation is used. A comma, followed by a subscripted index, denotes partial diferentiation with respect to the spatial coordinate denoted by that index, and a superposed dot indicates a time derivative. Both the stress tensor, σ_{ij}, and the infinitesimal strain sensor, ε_{kl}, are considered symmetric. Thus, the general constitutive relations for an anisotropic linear elastic material are given by

$$\sigma_{ij} = C_{ijkl}\varepsilon_{kl} \tag{3.3}$$

Here, the C_{ijkl} are the effective moduli of the composite, and the stresses and strains are understood to be the volume averages corresponding to a representative volume element defined within the composite continuum. Furthermore, the fourth-order elasticity tensor C_{ijkl} is considered to be positive definite, with 81 components, all of which are not independent. With the assumption of the existence of a strain energy density function and considering symmetry of the stress and strain tensors, the C_{ijkl} tensor satisfies the following relations:

$$C_{ijkl} = C_{jikl} = C_{ijlk} = C_{klij} \tag{3.4}$$

In view of the above symmetry relations, the number of independent components of the elasticity tensor for the most general anisotropic medium are reduced to 21. This number can be further reduced depending upon inherent material symmetries that can exist in the anisotropic material [48].

By combining Eqs. (3.1)–(3.3), the equation of wave motion becomes

$$C_{ijkl}u_{k,jl} = \rho\ddot{u}_i \tag{3.5}$$

Using a method outlined by Musgrave [49], the plane wave solution to Eq. (3.5) can be assumed as

$$u_i = A_i f(n_i x_i - c_n t) \tag{3.6}$$

where

A_i = components of the amplitude vector which are independent of time and spatial coordinates

n_i = direction cosines of the normal to the wave front

c_n = phase velocity

f = phase shape function, which is an appropriate differentiable function

The phase velocity c_n has the following relationship with respect to the circular frequency:

$$c_n = \frac{\omega}{r} = \frac{\omega\lambda}{2\pi} \tag{3.7}$$

where

ω = circular frequency

r = wave number

λ = wave length

To obtain the condition for solution of Eq. (3.6) and to satisfy the equation of motion (Eq. (3.5)), Eq. (3.6) is substituted into Eq. (3.5). This substitution gives

$$(C_{ijkl}n_j n_l - \rho c_n^2 \delta_{ik})A_k = 0 \tag{3.8}$$

The elastic quantities within the parentheses can be conveniently defined in terms of Christoffel stiffnesses as

$$\Gamma_{ik} = \Gamma_{ki} = C_{ijkl}n_jn_l \tag{3.9}$$

The condition for non-zero solutions of Eq. (3.8) can be written as

$$\det|\Gamma_{ki} - \rho c_n^2 \delta_{ik}| = 0 \tag{3.10}$$

This condition becomes an eigenvalue problem yielding a cubic equation in ρc_n^2 with the implication that there may be three velocities of propagation for given wave normals (n_1, n_2, n_3). Due to symmetry of the elasticity tensor C_{ijkl}, the Christoffel stiffness tensor Γ_{ik} is also symmetric and is positive definite. Thus, all the eigenvalues are real and if they are distinct, the corresponding three eigenvectors are mutually orthogonal. Therefore, it can be asserted that the characteristics of the elastic plane waves in an anisotropic material are fully determined by the elasticity tensor and the direction of wave propagation. It may be noted that, in general, the waves in an anisotropic medium are of mixed mode, that is, neither pure longitudinal nor pure transverse, except in cases when they travel along the axes of material symmetry.

By using contracted index notations [50], the Christoffel stiffnesses (Eq. (3.9)) can be written as

$$\Gamma_{11} = C_{11}n_1^2 + C_{66}n_2^2 + C_{55}n_3^2 + 2C_{16}n_1n_2 + 2C_{15}n_1n_3 + 2C_{56}n_2n_3$$

$$\Gamma_{12} = C_{16}n_1^2 + C_{26}n_2^2 + C_{45}n_3^2 + (C_{12} + C_{66})n_1n_2$$
$$\qquad + (C_{56} + C_{14})n_1n_3 + (C_{46} + C_{25})n_2n_3$$

$$\Gamma_{13} = C_{15}n_1^2 + C_{46}n_2^2 + C_{35}n_3^2 + (C_{56} + C_{14})n_1n_2$$
$$\qquad + (C_{13} + C_{55})n_1n_3 + (C_{35} + C_{36})n_2n_3$$

$$\Gamma_{22} = C_{66}n_1^2 + C_{22}n_2^2 + C_{44}n_3^2 + 2C_{26}n_1n_2 + 2C_{46}n_1n_3 + 2C_{24}n_2n_3$$

$$\Gamma_{23} = C_{56}n_1^2 + C_{24}n_2^2 + C_{34}n_3^2 + (C_{25} + C_{46})n_1n_2$$
$$\qquad + (C_{36} + C_{45})n_1n_3 + (C_{23} + C_{44})n_2n_3$$

$$\Gamma_{33} = C_{55}n_1^2 + C_{44}n_2^2 + C_{33}n_3^2 + 2C_{45}n_1n_2 + 2C_{35}n_1n_3 + 2C_{34}n_2n_3$$

$$\tag{3.11}$$

and the elastic stiffnesses can be written in the form

$$
\begin{Bmatrix} \sigma_{11} \\ \sigma_{22} \\ \sigma_{33} \\ \sigma_{23} \\ \sigma_{13} \\ \sigma_{12} \end{Bmatrix} = \begin{bmatrix} C_{11} & C_{12} & C_{13} & C_{14} & C_{15} & C_{16} \\ & C_{22} & C_{23} & C_{24} & C_{25} & C_{26} \\ & & C_{33} & C_{34} & C_{35} & C_{36} \\ & \text{symmetric} & & C_{44} & C_{45} & C_{46} \\ & & & & C_{55} & C_{56} \\ & & & & & C_{66} \end{bmatrix} \begin{Bmatrix} \varepsilon_{11} \\ \varepsilon_{22} \\ \varepsilon_{33} \\ 2\varepsilon_{23} \\ 2\varepsilon_{13} \\ 2\varepsilon_{12} \end{Bmatrix}
$$

$$(3.12)$$

3.3.1 Orthotropic Materials

An aligned fiber composite lamina can reasonably be assumed to possess three mutually perpendicular planes of elastic symmetry. The material symmetry axes x_1, x_2, and x_3, are shown in Figure 3.1 with the x_1 direction aligned along the fiber direction for a unidirectional composite lamina. Thus, the number of elastic constants for an orthotropic material are reduced to 9 from 21 as for the case of the most general anistropic material. The constitutive relation (Eq. 3.12) then reduces to

$$
\begin{Bmatrix} \sigma_{11} \\ \sigma_{22} \\ \sigma_{33} \\ \sigma_{23} \\ \sigma_{13} \\ \sigma_{12} \end{Bmatrix} = \begin{bmatrix} C_{11} & C_{12} & C_{13} & 0 & 0 & 0 \\ & C_{22} & C_{23} & 0 & 0 & 0 \\ & & C_{33} & 0 & 0 & 0 \\ & \text{symmetric} & & C_{44} & 0 & 0 \\ & & & & C_{55} & 0 \\ & & & & & C_{66} \end{bmatrix} \begin{Bmatrix} \varepsilon_{11} \\ \varepsilon_{22} \\ \varepsilon_{33} \\ 2\varepsilon_{23} \\ 2\varepsilon_{13} \\ 2\varepsilon_{12} \end{Bmatrix}
$$

$$(3.13)$$

and Eq. (3.11) reduces to

$$
\begin{aligned}
\Gamma_{11} &= C_{11}n_1^2 + C_{66}n_2^2 + C_{55}n_3^2 \\
\Gamma_{12} &= (C_{12} + C_{66})n_1 n_2 \\
\Gamma_{13} &= (C_{13} + C_{55})n_1 n_3 \\
\Gamma_{22} &= C_{66}n_1^2 + C_{22}n_2^2 + C_{44}n_3^2 \\
\Gamma_{23} &= (C_{23} + C_{44})n_2 n_3 \\
\Gamma_{33} &= C_{55}n_1^2 + C_{44}n_2^2 + C_{33}n_3^2
\end{aligned}
$$

$$(3.14)$$

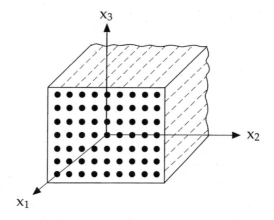

Figure 3.1 Coordinate system for unidirectional composite.

It is of interest to discuss the solution of Eq. (3.11) for the special case of orthotropic materials. For a plane wave traveling along the x_1 direction (Figure 3.1) $n_1 = 1$, and $n_2 = n_3 = 0$, Eq. (3.14) reduces to

$$\Gamma_{11} = C_{11}, \qquad \Gamma_{22} = C_{66}, \qquad \Gamma_{33} = C_{55}$$
$$\Gamma_{12} = \Gamma_{13} = \Gamma_{23} = 0 \tag{3.15}$$

Therefore, the characteristic roots of the eigen-equation (3.10) are given by

$$(C_{11} - \rho c_n^2)(C_{66} - \rho c_n^2)(C_{55} - \rho c_n^2) = 0 \tag{3.16}$$

With the following three distinct roots given as

$$c_{11L} = \sqrt{\frac{C_{11}}{\rho}}$$

$$c_{12T} = \sqrt{\frac{C_{66}}{\rho}} \tag{3.17}$$

$$c_{13T} = \sqrt{\frac{C_{55}}{\rho}}$$

The above development regarding a plane wave propagating along the x_1 direction yields an interesting result in that particle motion along three coordinate directions is excited, and each particle motion can be associated with a distinct wave speed. The fastest wave speed associated with a particle motion along the x_1 direction is called a longitudinal wave (c_{11L}) and waves associated with particle motions along the x_2 and x_3 directions are called the shear or transverse waves (c_{12T}, c_{13T}). The data contained in Table 3.1 summarize results for various wave types [50].

3.3.2   Transversely Isotropic Materials

If the fiber distribution within the cross-section shown in Figure 3.1 is uniform, then additional symmetry is imposed and the composite material can be considered to be transversely isotropic.

In this case, the number of independent elastic constants is reduced to five, since

$$C_{13} = C_{12}; \quad C_{22} = C_{23}; \quad C_{44} = C_{55} \tag{3.18}$$

3.3.3   Stress-Wave Velocity Measurements in Fiber-Reinforced Rods and Plates

Since the stress-wave velocities are directly related to the corresponding elastic constants, wave velocity measurements can be used to determine these constants. A significant amount of literature is available that deals primarily with these measurements as, for example, for various crystal systems possessing all possible elastic symmetries.

These types of measurements have been extended to the study of wave motion in various composite structural elements. The following development considers wave velocity measurements in filamentary reinforced rods and thin plates as an illustration of the wave motions in these new structural materials.

Rods   The structural configuration to be termed as "rod" is basically a one-dimensional element in which the transverse dimensions relative to the longitudinal axis are small enough for the wave motion assumed to be one-dimensional propagating along the longitudinal axis. The rod can be reinforced in two ways: (1) when the continuous fibers are uniformly aligned along the longitudinal axis; and (2) when

TABLE 3.1 Relations Between Elastic Constants and Wave Velocities for an Orthotropic Elastic Material [50,51]*

Wave Normal	Particle Direction	Wave Type	Phase Velocity Relation
$n_1 = 1$	x_1	Longitudinal	$\rho c_{11L} = C_{11}$
$n_2 = 0$	x_2	Transverse	$\rho c_{12T} = C_{66}$
$n_3 = 0$	x_3	Transverse	$\rho c_{13T} = C_{55}$
$n_1 = 0$	x_1	Transverse	$\rho c_{21T} = C_{66}$
$n_2 = 1$	x_2	Longitudinal	$\rho c_{22L} = C_{22}$
$n_3 = 0$	x_3	Transverse	$\rho c_{23T} = C_{44}$
$n_1 = 0$	x_1	Transverse	$\rho c_{31T} = C_{55}$
$n_2 = 0$	x_2	Transverse	$\rho c_{32T} = C_{44}$
$n_3 = 1$	x_3	Longitudinal	$\rho c_{33L} = C_{33}$
$n_1 = 0$	x_1	Quasi-transverse	$\rho c_n^2 = C_{66}n_2^2 + C_{55}n_3^2$
$n_2 \neq 0$	x_2–x_3	Quasi-longitudinal	$(C_{22}n_2^2 + C_{44}n_3^2 - \rho c_n^2)\,(C_{44}n_2^2 + C_{33}n_3^2 - \rho c_n^2)$
$n_3 \neq 0$	Plane	Quasi-transverse	$= (C_{23} + C_{44})^2\, n_2^2 n_3^2$
$n_2 = 0$	x_3	Transverse	$\rho c_n^2 = C_{66}n_1^2 + C_{44}n_3^2$
$n_1 \neq 0$	x_1–x_3	Quasi-longitudinal	$(C_{11}n_1^2 + C_{55}n_3^2 - \rho c_n^2)\,(C_{55}n_1^2 + C_{33}n_3^2 - \rho c_n^2)$
$n_3 \neq 0$	Plane	Quasi-transverse	$= (C_{13} + C_{55})^2\, n_1^2 n_3^2$
$n_3 = 0$	x_3	Transverse	$\rho c_n^2 = C_{55}n_1^2 + C_{44}n_2^2$
$n_1 \neq 0$	x_1–x_2	Quasi-longitudinal	$(C_{11}n_1^2 + C_{66}n_2^2 - \rho c_n^2)\,(C_{66}n_1^2 + C_{22}n_2^2 - \rho c_n^2)$
$n_2 \neq 0$	Plane	Quasi-transverse	$= (C_{12} + C_{66})^2\, n_1^2 n_2^2$

* Reprinted with permission from the Vibration Institute.

the fibers are uniformly oriented in a transverse spacing, that is perpendicular to the longitudinal direction. In both these cases, the wave propagation velocity along the axis of the rod is given by

$$c_{1L} = \sqrt{\frac{C_{11}}{\rho_c}} = \sqrt{\frac{E_{1L}}{\rho_c}} \tag{3.19}$$

where E_{1L} is modulus of elasticity measured along the rod axis, and ρ_c is the density of the rod. Since the rod consists of fibers and matrix materials, the elastic modulus and density will depend upon the elastic moduli, densities, and volume fractions.

In the case when the fibers are aligned along the rod axis (Figure 3.2a), the "effective modulus" E_{1L} and "effective density" ρ_c can be obtained using the rule of mixtures, given by

$$E_{1L} = E_f V_f + E_m V_m$$
$$\rho_c = \rho_f V_f + \rho_m V_m \tag{3.20}$$
$$V_f + V_m = 1$$

where

$V_f, \ V_m$ = volume fraction of fibers and matrix, respectively
$E_f, \ E_m$ = elastic modulus of fibers and matrix, respectively

Therefore, the longitudinal wave speed in the rod along the fibers is given by

$$c_{1L} = \left[\frac{E_f V_f + E_m V_m}{\rho_f V_f + \rho_m V_m}\right]^{1/2} \tag{3.21}$$

In the case where the fibers are aligned normal to the rod axis (Figure 3.2b), the effective modulus can be estimated to be [50]

$$E_{2L} = \frac{E_m(1 + 2\eta V_f)}{1 - \eta V_f}$$

where

$$\eta = \frac{(E_f/E_m) - 1}{(E_f/E_m) + 2} \tag{3.22}$$

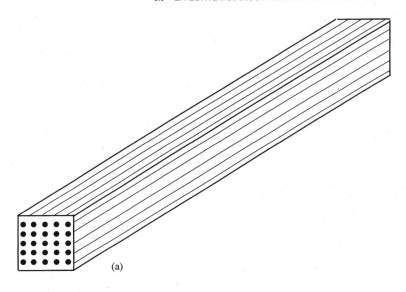

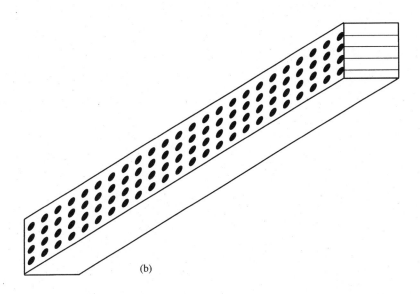

Figure 3.2 Uniaxial rod with longitudinal (a) and transverse (b) filaments.

The longitudinal velocity of a stress wave traveling along the rod axis but transverse to the fibers is given by

$$C_{2L} = \left[\frac{E_m(1 + 2\eta V_f)}{(\rho_f V_f + \rho_m V_m)(1 - \eta V_f)}\right] \tag{3.23}$$

Experimental results obtained for the case of steel filament/epoxy matrix long rod specimens have been compared with Eqs. (3.22) and (3.23) by Ross and Sierakowski [50], and are shown in Figure 3.3. It is interesting to observe that the wave speed along the rod axis with fibers oriented in the transverse direction remains below that of the pure matrix for volume fractions of steel filaments up to approximately 0.7.

Composite Plates   For a unidirectional fiber-reinforced composite plate, as shown in Figure 3.4, the axes x_1, x_2, x_3, with the axis x_1 oriented along the fiber direction, are the principal material axes. With the assumption of a uniform distribution of fibers, the lamina can be considered to be transversely isotropic with the plane perpendicular to the fiber direction as the plane of isotropy. Further, if it is assumed that the plate is thin and that the wave length is much larger than the thickness of the plate, then the wave propagation velocities along the two in-plane principal directions can be given by [52,53]

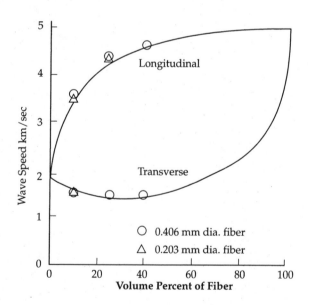

Figure 3.3   Wave speed for steel filament epoxy matrix long rod specimens [50]. (Reprinted with permission.)

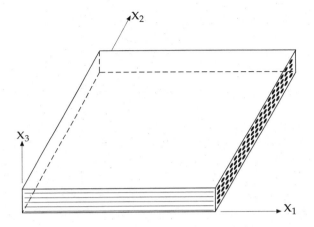

Figure 3.4 Uniaxial composite plate with axis x_3 normal to the plane of plate.

$$c_{11L} = \sqrt{\frac{C_{11}}{\rho}} = \sqrt{\frac{E_{11}}{\rho(1 - \nu_{12}\nu_{21})}}$$

$$c_{22L} = \sqrt{\frac{C_{22}}{\rho}} = \sqrt{\frac{E_{22}}{\rho(1 - \nu_{12}\nu_{21})}}$$

(3.24)

The primary flexural wave velocities with particle motions normal to the plates are given by

$$c_{13T} = \sqrt{\frac{C_{55}}{\rho}} = \sqrt{\frac{G_{13}}{\rho}} = \sqrt{\frac{G_{12}}{\rho}}$$

$$c_{23T} = \sqrt{\frac{C_{44}}{\rho}} = \sqrt{\frac{G_{23}}{\rho}} = \sqrt{\frac{E_{33}}{2\rho(1 + \nu_{23})}}$$

(3.25)

$$= \sqrt{\frac{E_{22}}{2\rho(1 + \nu_{23})}}$$

There are also transverse waves associated with the transverse shear deformations. The flexural wave propagating at 45° with $n_1 = n_2 = 1/\sqrt{2}$, $n_3 = 0$ and the particle motion along the x_3 direction (normal to the plate) travel at a speed given by

$$c_{(45)\text{T}} = \sqrt{\frac{C_{44} + C_{55}}{2\rho}} = \sqrt{\frac{G_{13} + G_{23}}{2\rho}}$$

$$= \sqrt{\frac{c_{13\text{T}}^2 + c_{23\text{T}}^2}{2}}$$

(3.26)

Experimental measurements of these wave speeds for a lamina as well as for laminated composites under impact loading have been determined by a number of researchers such as Daniel and Liber [52] and Takeda [53]. Further discussion is included in the following sections.

3.4 WAVE MOTIONS IN COMPOSITE PLATES UNDER IMPACT LOADING

A number of theoretical and experimental studies on the wave motion in composite plates under impact loading have been reviewed by Moon [3]. Investigations involving transverse impact loadings are of practical interest and are considered in this chapter. It has been shown that the wave motion in composite plates under transverse impact is composed of five waves, two of which are related to in-plane motion and three related to flexural plate motions. The types of waves, their velocities, and their effect within the plate are discussed using an approach suggested by Moon [3].

Consider a composite plate made of unidirectional plies oriented at angles $\pm\phi$ from the symmetric axis with a stacking sequence such that bending-extensional coupling does not exist. Furthermore, let the wavelengths involved be greater than 100 fiber diameters and greater than the ply-thickness so that an effective modulus theory can be used. Thus, the laminated plate consisting of n plies can be treated as an equivalent anisotropic plate with the principal elastic properties calculated from the static analysis of the n-ply plate [50].

The basic equations of motion for such anisotropic plates are given by Eq. (3.6). For a plate element of total thickness $2b$, as shown in Figure 3.5, the displacement components can be expanded in Legendre polynomials in the thickness direction as discussed by Moon [54].

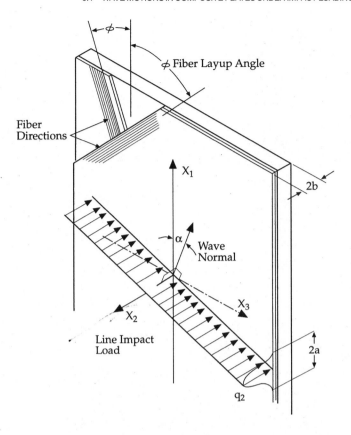

Figure 3.5 Diagram of transversely impacted composite plate geometry [56].

$$u_i = \sum p_n(h)u_i^{(n)}(x_1, x_3, t) \qquad (3.27)$$

where the thickness variable

$$h = x_2/b$$
$$p_0(h) = 1$$

$$p_1(h) = h, \quad \text{and} \quad p_2(h) = \frac{3h^2 - 1}{2}$$

The quantities $u_i^{(n)}$ have the following meanings:

u_1^0, u_3^0 = in-plane deformation

u_2^0 = transverse displacement of the plate

u_1^1, u_3^1 = measures of the bending strains

u_2^1 = measure of stretching along thickness direction of the plate

Alternatively, a variational problem can be solved instead of directly solving Eq. (3.5). These equations can be integrated through the plate thickness as

$$\int_A \int_{-1}^{1} (\sigma_{ij,j} - \rho \ddot{u}_i)\delta u_i b \ dh \ dA = 0 \qquad (3.28)$$

here, δu_i and σ_{ij} can be evaluated using Eq. (3.27). Equation (3.28) leads to

$$b\sigma_{\beta i,\beta}^{(n)} + [p_n(h)\sigma_{2i}]_{-1}^{1} - \sigma_{2i}^{(n)} = \rho b \frac{2}{2h+1} \ddot{u}_i^{(n)} \qquad (3.29)$$

where $\beta = 1, 3$

$$\sigma_{\beta i}^{(n)} = \int_{-1}^{1} p_n(h)\sigma_{\beta i} \ dh$$

$$\sigma_{2i}^{(n)} = \int_{-1}^{1} \frac{dp_n}{dh} \sigma_{2i} h \qquad (3.30)$$

With the impact force $(-q_2)$ applied to the top surface of the plate, the boundary conditions can be written as

$$\sigma_{22}(h = 1) = -q_2$$
$$\sigma_{22}(h = -1) = \sigma_{21}(h = \pm 1) = \sigma_{23}(h = \pm 1) = 0 \qquad (3.31)$$

The infinite set of equations can be truncated by dropping out all the higher-order terms, such as $u_1^{(2)}$, $u_3^{(2)}$, and so forth. Furthermore, for orthogonal symmetry it is assumed that $n = 0, 1,$ and 2, and some terms containing second derivatives for $n = 2$ in Eq. (3.29) are also dropped. Physically, this implies that higher-frequency terms have been ignored. The terms remaining are the following: $u_1^0, u_2^0, u_3^0, u_1^1, u_2^1, u_3^1,$ and $u_2^{(2)}$. Based on these assumptions, all strain components except for ε_{12} and ε_{23} are, thus, linear functions of the thickness variable h.

By expanding the displacement relations (Eq. 3.27) and using the constitutive relations (Eq. (3.13)) and boundary conditions (Eq. (3.31)) at $h = 1$, and $h = -1$, the following relations can be obtained.

$$C_{12}\left(\frac{\partial u_1^0}{\partial x_1} + \frac{\partial u_1^1}{\partial x_1}\right) + C_{22}\left(\frac{u_2^1}{b} + \frac{3u_2^2}{b}\right) + C_{23}\left(\frac{\partial u_3^0}{\partial x_3} + \frac{\partial u_3^1}{\partial x_3}\right) = q_2$$

$$C_{12}\left(\frac{\partial u_1^0}{\partial x_1} - \frac{\partial u_1^1}{\partial x_1}\right) + C_{22}\left(\frac{u_2^1}{b} - \frac{3u_2^2}{b}\right) + C_{23}\left(\frac{\partial u_3^0}{\partial x_3} - \frac{\partial u_3^1}{\partial x_3}\right) = 0$$

(3.32)

By adding together and subtracting each of these equations, equations for the bending strains u_2^1 and u_2^2 can be obtained as follows:

$$q_2 - 2\left[C_{22}\frac{u_2^1}{b} + C_{12}\frac{\partial u_1^0}{\partial x_1} + C_{23}\frac{\partial u_3^0}{\partial x_3}\right] = 0$$

$$q_2 - 2\left[C_{22}\frac{3u_2^{(2)}}{b} + C_{12}\frac{\partial u_1^1}{\partial x_1} + C_{23}\frac{\partial u_3^1}{\partial x_3}\right] = 0$$

(3.33)

Using these relations, the terms u_2^1, and $u_2^{(2)}$ can be eliminated from Eq. (3.29) to yield the following wave equations:

$$\rho\frac{\partial^2 u_1^0}{\partial t^2} = C_{11}^*\frac{\partial^2 u_1^0}{\partial x_1^2} + C_{55}\frac{\partial^2 u_1^0}{\partial x_3^2} + (C_{55} + C_{13}^*)\frac{\partial^2 u_3^0}{\partial x_1 \partial x_3} + \frac{C_{12}}{2C_{22}}\frac{\partial q_2}{\partial x_1}$$

(3.34)

$$\rho\frac{\partial^2 u_3^0}{\partial t^2} = C_{33}^*\frac{\partial^2 u_3^0}{\partial x_3^2} + C_{55}\frac{\partial^2 u_3^0}{\partial x_1^2} + (C_{55} + C_{13}^*)\frac{\partial^2 u_1^0}{\partial x_1 \partial x_3} + \frac{C_{23}}{2C_{22}}\frac{\partial q_2}{\partial x_3}$$

(3.35)

$$\rho\frac{\partial^2 u_2^0}{\partial t^2} = C_{66}\frac{\partial^2 u_2^0}{\partial x_1^2} + C_{44}\frac{\partial^2 u_2^0}{\partial x_3^2} + C_{66}\frac{1}{b}\frac{\partial u_1^1}{\partial x_1} + C_{44}\frac{1}{b}\frac{\partial u_3^1}{\partial x_3} + \frac{1}{2b}q_2$$

(3.36)

$$\rho\frac{\partial^2 u_1^1}{\partial t^2} = C_{11}^*\frac{\partial^2 u_1^1}{\partial x_1^2} + C_{55}\frac{\partial^2 u_1^1}{\partial x_3^2} + (C_{55} + C_{13}^*)\frac{\partial^2 u_3^1}{\partial x_1 \partial x_3}$$

$$- \frac{3}{b}C_{66}\left(\frac{\partial u_2^0}{\partial x_1} + \frac{u_1^1}{b}\right) + \frac{C_{12}}{2C_{22}}\frac{\partial q_2}{\partial x_1}$$

(3.37)

$$\rho \frac{\partial^2 u_3^1}{\partial t^2} = C_{33}^* \frac{\partial^2 u_3^1}{\partial x_3^2} + C_{55} \frac{\partial^2 u_3^1}{\partial x_1^2} + (C_{55} + C_{13}^*) \frac{\partial^2 u_1^1}{\partial x_1 \partial x_3}$$

$$- \frac{3}{b} C_{44} \left(\frac{\partial u_2^0}{\partial x_3} + \frac{u_3^1}{b} \right) + \frac{C_{23}}{2 C_{22}} \frac{\partial q_2}{\partial x_3} \qquad (3.38)$$

where

$$C_{11}^* = C_{11} - \frac{C_{12}^2}{C_{22}}$$

$$C_{33}^* = C_{33} - \frac{C_{23}^2}{C_{22}}$$

$$C_{13}^* = C_{13} - \frac{C_{12} C_{23}}{C_{22}}$$

Equations (3.34) and (3.35) govern the in-plane wave motions, while Eqs. (3.36)–(3.38) govern the flexural wave motions. It is interesting to observe that a transverse force can generate both in-plane and flexural waves. Furthermore, these waves, as discussed later, can be related to damage propagation in laminated plates.

As a first step, consider in-plane motions (Eqs. (3.34) and (3.35)) and use solutions of the form given below, as deduced from the general solution represented by Eq. (3.6).

$$u_1^0 = A_1 f(n_i x_i - c_n t)$$

$$u_3^0 = A_3 f(n_i x_i - c_n t) \qquad (3.39)$$

By substituting the above solution into the governing Eqs. (3.34) and (3.35), we get the following algebraic equation for a given normal n_i:

$$\begin{bmatrix} \Gamma_{11}^* & \Gamma_{13}^* \\ \Gamma_{31}^* & \Gamma_{33}^* \end{bmatrix} \begin{bmatrix} A_1 \\ A_3 \end{bmatrix} = c_n^2 \begin{bmatrix} A_1 \\ A_3 \end{bmatrix} \qquad (3.40)$$

where

$$\Gamma_{11}^* = C_{11}^* n_1^2 + C_{55} n_3^2$$

$$\Gamma_{33}^* = C_{33}^* n_3^2 + C_{55} n_1^2$$

$$\Gamma_{13}^* = \Gamma_{31}^* = (C_{13}^* + C_{55}) n_1 n_3$$

$$n_1 = \cos \alpha, \qquad n_3 = \sin \alpha$$

The Γ_{ij}^* being positive definite ensures that two positive roots, C_{n1}^2 and C_{n2}^2, exist, each of them corresponding to the two in-plane waves for a given wave normal n_i. These roots are determined from the equation

$$\det[\Gamma_{ij}^* - \rho c_n^2 \delta_{ij}] = 0 \qquad (3.41)$$

For a displacement direction parallel to the wave normal, the wave is called longitudinal and for the displacement normal to n_i, the wave is called transverse. For isotropic materials, the larger root corresponds to the longitudinal wave and the smaller root to the transverse wave.

For flexural wave motions represented by Eqs. (3.36)–(3.38), it can be shown that plane wave solutions of the form Eq. (3.39) are harmonic functions which can be represented as discussed in reference [54].

$$\begin{bmatrix} u_1^1 \\ u_3^1 \\ u_2^0 \end{bmatrix} = \begin{bmatrix} -b\psi_3 \\ b\psi_1 \\ A_2 \end{bmatrix} e^{ik(n_i x_i - c_n t)} \qquad (3.42)$$

where ψ_1 and ψ_3 are the slopes of the plate mid-surface due to bending. The phase velocity c_n depends on the wave normal n_i as well as on the frequency $\omega = kc_n$, with dependence for various material anisotropies discussed by Mindlin [55].

Solution procedures for determining bending waves are based upon the following assumptions:

1. The displacement and stresses are continuous across the wave front.
2. The second derivatives of A_i are discontinuous across the wave front.
3. Jump discontinuities in the function $\psi(x_1, x_3)$ exist across the wave front.

Therefore,

$$[u_{2,1}^0] = [u_{2,3}^0] = 0$$
$$[u_{1,1}^1] = [u_{1,3}^1] = 0 \qquad (3.43)$$
$$[u_{3,1}^1] = [u_{3,3}^1] = 0$$

The equations of motion (Eqs. (3.36)–(3.38)) are valid on both sides of the wave front. These relations can be simplified to obtain

$$\rho\left[\frac{\partial^2 u_2^0}{\partial t^2}\right] = C_{66}\left[\frac{\partial^2 u_2^0}{\partial x_1^2}\right] + C_{44}\left[\frac{\partial^2 u_2^0}{\partial x_3^2}\right]$$

$$\rho\left[\frac{\partial^2 u_1^1}{\partial t^2}\right] = C_{11}^*\left[\frac{\partial^2 u_1^1}{\partial x_1^2}\right] + C_{55}\left[\frac{\partial^2 u_1^1}{\partial x_3^2}\right] + (C_{55}+C_{13}^*)\left[\frac{\partial^2 u_3^1}{\partial x_1 \partial x_3}\right] \qquad (3.44)$$

$$\rho\left[\frac{\partial^2 u_3^1}{\partial t^2}\right] = C_{33}^*\left[\frac{\partial^2 u_3^1}{\partial x_3^2}\right] + C_{55}^*\left[\frac{\partial^2 u_3^1}{\partial x_1^2}\right] + (C_{55}+C_{13}^*)\left[\frac{\partial^2 u_1^1}{\partial x_1 \partial x_3}\right]$$

By considering a jump in the acceleration and consequently in the strain for a plane wave front with unit normal n_i, the following relation can be obtained

$$\psi_{,ij} = \frac{n_i n_j}{c_n^2}\ddot{\psi} \qquad (3.45)$$

By using Eq. (3.45) in Eq. (3.44), the following linear algebraic relations between the discontinuities in acceleration across the wave front can be obtained:

$$\rho c_n^2 = C_{66}n_1^2 + C_{44}n_3^2$$

$$\begin{bmatrix} \Gamma_{11}^* & \Gamma_{13}^* \\ \Gamma_{13}^* & \Gamma_{33}^* \end{bmatrix}\begin{bmatrix} \dfrac{\partial^2 u_1^1}{\partial t^2} \\ \dfrac{\partial^2 u_3^1}{\partial t^2} \end{bmatrix} = c_n^2\begin{bmatrix} \dfrac{\partial^2 u_1^1}{\partial t^2} \\ \dfrac{\partial^2 u_3^1}{\partial t^2} \end{bmatrix} \qquad (3.46)$$

where the Γ_{ij}^* are given by Eq. (3.40).

This result is interesting since the above relation indicates that the jump in the bending accelerations $\partial^2 u_1^1/\partial t^2$ and $\partial^2 u_3^1/\partial t^2$ travel at the same speed as the in-plane motion. For the corresponding wave front associated with the jump in acceleration $\partial^2 u_2^0/\partial t^2$ (the first equation of (3.44)) and for a mid-plane symmetric laminate (i.e., $C_{66} = C_{44}$), one recovers the situation where the bending wave becomes directionally isotropic.

Based upon the theoretical developments discussed above, Moon [56] proceeded to solve two types of impact problems. The first was that of a transverse line load and the second that of a concentrated central impact loading distributed over a small area. The material considered was graphite epoxy with ply orientations of $0°$, $\pm15°$, $\pm30°$, and $\pm45°$, respectively, for the laminated plates studied. The primary quantities of interest involved the determination of transient distributions of membrane, flexural, and interlaminar stresses. These stresses have been found to play an important role in the initiation and propagation of various damage modes (e.g., matrix cracking, fiber fracture, delaminations) in laminated composites. Important elements in the solution procedure include estimation of the Hertzian contact pressure over the impacted area, and the introduction of Fourier transforms for determining the required displacements, strains, and stresses. Also, the equivalent elastic constants for various graphite-epoxy ply lay-ups were calculated using the procedures proposed by Chamis [57]. As a means to illustrate the use of analysis discussed above the problem of a central impact is briefly discussed. For additional details, the reader is referred to the original reference report [56].

The primary steps taken for obtaining solutions of the governing Eqs. (3.34)–(3.38) is to first prescribe the loading function $q_2(x_1, x_3, t)$, then to introduce Fourier transform variables for transforming the original equations, and finally, apply inversion procedures to return to the original plane of interest.

In order to determine the stresses due to impact, a Hertzian impact problem is taken as a first step. In this approach, the pressure distribution due to impact on a composite plate initially at rest is assumed to be [56],

$$P = P_0 \left[1 - \left(\frac{r}{a}\right)^2 \right]^{1/2} \tag{3.47}$$

where

$$P_0 = \tfrac{3}{2} F_0 / \pi a^2$$

$$F_0 = 3.36 \frac{MV}{\tau}$$

$$\tau = \frac{2.943 \alpha_1}{V}$$

$$\alpha_1 = \left[\frac{5}{4} \frac{MV^2}{k_2} \right]^{2/5}$$

$$k_2 = \tfrac{4}{3} R^{1/2} \left[\frac{1 - \nu_1^2}{E_1} + \frac{1}{C_{22}} \right]^{-1}$$

$E_1, \nu_1 =$ elastic moduli of impacting sphere

$C_{22} =$ elastic modulus of composite plate

$M, R =$ mass and radius of impacting sphere

$V =$ instantaneous velocity of sphere

$\tau =$ contact time

$a = \sqrt{\alpha_1 R}$

It should be noted that the above description of contact force, impact time, and contact pressure is based upon static indentation tests, and is thus approximate. For dynamic impacts subject to strike velocities in the range of 100–500 m/s, the above estimation can be in serious error. However, based upon the above estimate of contact force, the stresses calculated include the average membrane stresses $(\sigma_{11}^0 + \sigma_{33}^0)/2$, the average bending stresses $(\sigma_{11}^1 + \sigma_{33}^1)/2$, and the maximum interlaminar shear stress $(\sigma_{21}^2 + \sigma_{23}^2)/2$. The impact load distribution shown in Figure 3.6, and the flexural stress distribution for all laminates considered have been shown in Figure 3.7. Results which can be drawn from this study include the following:

1. Stress levels are generally largest along the fiber directions, with the maximum stresses propagating with the slowest flexural wave. The $\pm 15°$ lay-up produces lower flexural stresses than do the 0, $\pm 30°$ or $\pm 45°$ lay-ups. The flexural stresses appear to depend upon the ratio of impactor nose radius to plate thickness.

2. The average membrane stresses immediately after impact, due to in-plane wave motion, appear to be smaller for smaller fiber lay-up angles.

3. The interlaminar shear stresses on the interlaminar plane appear to be insensitive to lay-up angle.

3.5 EXPERIMENTAL WAVE PROPAGATION STUDIES UNDER IMPACT

Experimental investigations for impact loaded structural materials can be broadly classified into three categories based upon wave shape and amplitude [3], these being:

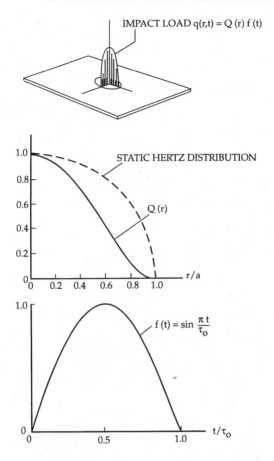

Figure 3.6 Impact load distribution [56].

1. low-amplitude sinusoidal waves or ultrasonic waves;
2. low-amplitude transient waves; and
3. high-amplitude transient waves.

Several experimental methods have been developed to study the wave motion in various composite systems. These methods include the generation and monitoring of waves under each of the above categories. Ultrasonics, for example, has been used successfully, for monitoring and characterizing the damage in composites. Techniques based upon these waves form an important part of a separate field of

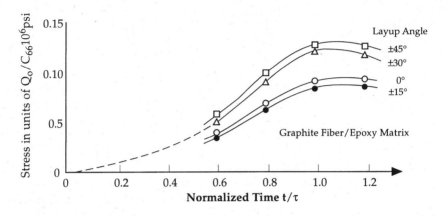

Figure 3.7 Variation of flexural stresses with time for angle-ply laminates. Flexural stress $(t_{11} + t_{33})/2$ at origin vs. time, $a/b = 1.0$. Normalized contact time $\tau V_3/b = 1.0$, $V_3 = 1.18\text{mm}/\mu s$ [56].

study known as non-destructive evaluative (NDE) techniques. The experimental techniques developed to introduce low-amplitude transient waves in composites include mechanical strikers, small charge explosives, and shock tubes. For generating high-amplitude waves, exploding foils, higher-order explosives, and flyer plate type devices are more popular. Some of these techniques have already been discussed in detail in Chapter 2.

Daniel and Liber [52] made an interesting study of the wave propagation in unidirectional and angle-ply $[0_2 \pm 45]_{2s}$ composite laminates made of graphite-epoxy and boron-epoxy. Typical plate specimens were subjected to a low-velocity impact by silicon rubber spheres of 7.9 mm diameter. The impact velocities were kept specifically low, up to 250 m/s, in order to avoid the introduction of any potential damage in the specimens. Strain-gage signals were monitored on these specimens by mounting strain gages on the surface, and also at the interlaminar planes through the thickness. The transient records were analyzed to determine wave characteristics, that is, the types of wave and their propagation velocities, and other quantities such as peak strains, strain rates, and attenuation characteristics.

As discussed, the principal wave motions under transverse impact loading consist primarily of five waves: two waves related to in-plane motion and three related to flexural deformations. The measured and calculated wave velocities for the various types of waves have been included in Tables 3.2 and 3.3 [52]. The following significant observations can be made from these tables:

1. The longitudinal in-plane wave along the fiber direction (c_{11LI}) in a unidirectional laminate (0_{16}) travels, as expected, faster than any other in-plane or flexural wave.
2. There is a noticeable discrepancy between measured and computed values of wave velocities for both types of laminates. Several reasons have been suggested for this discrepancy. The measured value of c_{11LI} may not be as accurate as expected, due in part to the fact that the measured time interval (being extremely short) is not quite accurate. The calculated value of c_{22LI} is lower than the measured value, most likely due to the

TABLE 3.2   Wave Propagation Velocities in Transversely Impacted (0_{16}) Boron/Epoxy Specimen [52]

Velocity Direction and Type of Wave	Velocity, m/s (in/s)	
	Measured	Calculated
c_{11LI}	12 700 (500 000)	10 110 (398 100)
c_{22LI}	3 640 (143 300)	3 280 (129 000)
c_{45LI}	3 380 (133 200)	3 380 (133 000)
c_{11LF}	2 120 (83 630)	1 650 (65 000)
c_{22LF}	1 750 (68 790)	1 890 (74 500)
c_{45LF}	1 940 (76 300)	1 890 (74 500)

TABLE 3.3   Wave Propagation Velocities in Normally Impacted ($0_2/45)_{2s}$ Boron/Epoxy Specimen [52]

Velocity Direction and Type of Wave	Velocity, m/s (in/s)	
	Measured	Calculated
c_{xxLI}	3 520 (138 400)	3 310 (130 200)
c_{yyLI}	6 970 (274 400)	7 750 (305 000)
c_{45LI}	6 230 (245 200)	5 850 (230 000)
c_{xxLF}	1 810 (71 400)	1 650 (65 000)
c_{yyLF}	1 900 (74 800)	1 650 (65 000)
c_{45LF}	1 770 (69 600)	1 650 (65 000)

value of the static transverse elastic value E_2 used in the calculation which is lower than the dynamic value for the given impact loading situation. Similar discrepancies in flexural wave velocities are attributed to inaccuracies in the shear moduli G_{12}, G_{13} and G_{23}.

Further transient wave motion studies due to localized impact on composite laminated plates by Takeda [53] and Takeda et al. [58] are worth describing because of their correlations with transverse matrix cracking and delamination type of damage modes. These studies used cylindrical steel impactors of $\frac{3}{8}$ in in diameter and 1 in length with blunt ends impacting at the center of glass-epoxy cross-ply $((0)_5/(90)_5/(0)_5)$ and angle-ply $((+30)_5/(-30)_5/(+30)_5)$ laminated plates with a plate geometry of 6 in $\times$ 6 in and clamped along their four edges. These plate specimens were impacted at velocities ranging between 30 and 40 m/s. The wave motions within the specimens were measured by instrumented surface and imbedded strain gages with several strain-gage layouts used to monitor wave propagation of all the in-plane and flexural waves outside the region of the impact area.

A typical gage layout for a cross-ply $((0)_5/(90)_5/(0)_5)$ glass-epoxy laminated plate is shown in Figure 3.8, with the corresponding tran-

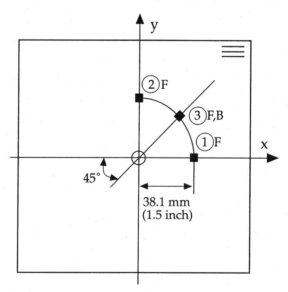

Figure 3.8 Surface-gage layout for $[(0°)_5/(90°)_5/(0°)_5]$ fiberglass\epoxy laminates. F and B denote front and back gages [53].

sient strains obtained from gages located at a distance of 1.5 in from the impact point shown in Figs. 3.9(a,b). The predominant wave in the 45° direction with respect to the x and y axis appears to be a flexural one (refer to gage records 3f and 3b which were obtained from the front and back gages, respectively). Notice that gages 3F and 3B are at the same location but on opposite sides of the plate; hence, the strain waves are symmetric to each other and the summation of their values is very close to zero. This indicates that the plate is

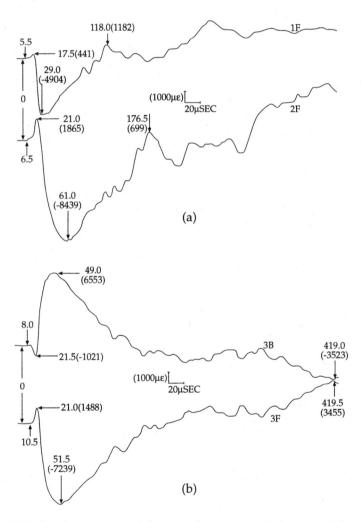

Figure 3.9 Strain-gage output from surface gages with layout in Figure 3.8. Impact velocity = 31.5 m/s [53].

in flexural motion during the impact, and that membrane effects are negligible.

The transverse matrix cracks on the impacted surface are believed to be generated by the tensile stresses due to the leading part of the transient flexural wave generated immediately after impact. The authors [53, 58] have suggested the following sequence of transient wave motions in a cross-ply composite plate under a central impact loading. First, an in-plane tensile wave arrives at the strain gages just away from the impact point, then, on the front (back) surface, a tensile (compressive) component of an oncoming flexural wave arrives followed by its large compressive (tensile) component. The largest amplitude component within the flexural wave may be the primary contributing factor to the delamination fracture observed in the composite laminates. The above conclusions have also been supported by Liu and Malvern [59] for glass-epoxy composite plates.

3.6   REFERENCES

1. Lee, T.C. and Ting, T.C.T. (1978) *A Review of Dynamic Response of Composites*, AMMRC TR 78-20, Dept. of Materials Engineering, Univ. of Illinois at Chicago Circle, April.

2. Lee, E.H. (1972) *Dynamics of Composite Materials*, ASME Applied Mechanics Division Series, Book No. H0078.

3. Moon, F.C. (1973) *A Critical Survey of Wave Propagation and Impact in Composite Materials*, Report No. 1103, Dept. of Aerospace & Mechanical Science, Princeton University.

4. Bedford, A., Drumheller, D.S., and Sutherland, H.J. (1976) "On modeling the dynamics of composite materials," *Mechanics Today*, **3**, 1–54; AMR 30 (1977), Rev. 3897.

5. Ting, T.C.T. (1986) "Dynamic response of composites," *Appl. Mech. Rev.*, **33**, 1629–1635, December 1980; update to "Dynamic response of composites," *Appl. Mech.* update, p. 349.

6. Gibson, R.F. (1990) "Dynamic mechanical properties of advanced composite materials and structures: a review of recent research," *The Shock and Vibration Digest*, **22**(8), 3–12.

7. Postma, G.W. (1955) "Wave propagation in a stratified medium," *Geophysics*, **20**, 780–806.

8. White, J.E. and Angona, F.A. (1955) "Elastic wave velocities in laminated media," *J. Acoustic Soc. America*, **27**, 311–317.

9. Weitsman, Y. (1972) "On wave propagation and energy scattering in materials reinforced by inextensible fibers," *Int. J. Solids & Structures*, **8**, 627–650.

10. Behrens, E. (1967) "Sound propagation in lamellar composite materials & advanced elastic constants," *J. Acoustic Soc. America*, **42**, 378.

11. Aboudi, J. and Weitsman, Y. (1972) "Impact deflection by oblique fibers in sparsely reinforced composites," *ZAMP*, **24**, 828–844.

12. Sun, C.T., Feng, W.H., and Koh, S.L. (1974) "A theory for physically non-linear elastic fiber-reinforced composites," *Int. J. Eng. Sci.*, **12**, 919–935.

13. Sun, C.T., Achenbach, J.D., and Herrmann, G. (1968) "Continuum theory of a laminated medium," *J. Appl. Mech.*, **35**, 467–475.

14. Achenbach, J.D., Sun, C.T., and Herrmann, G. (1968) "On the vibration of a laminated body," *J. Appl. Mech.*, **35**, 689–696.

15. Grot, R.A. and Achenbach, J.D. (1970) "Linear anisothermal theory for a viscoelastic laminated composite," *Acta Mech.*, **9**, 245–263.

16. Grot, R.A. and Achenbach, J.D. (1970) "Large deformations of a laminated composite," *Int. J. Solids Structures*, **6**, 641–659.

17. Grot, R.A. (1972) "A continuum model for curvilinear laminated composites," *Int. J. Solids Struct.*, **8**, 439–462.

18. Drumheller, D.S. and Sutherland, H.J. (1973) "A lattice model for composite materials," *J. Appl. Mech.*, **40**, 149–154.

19. Rausch, P.J. (1971) *Shock Formation and Pulse Attenuation in a Nonlinear Geometrically Dispersive Solid*, The Aerospace Corp., San Bernardino, CA, TR-0059 (S6816-75)-1, February.

20. Mindlin, R.D. (1964) "Microstructure in linear elasticity," *Arch. Rat. Mech. Anal.*, **16**, 51.

21. Ben-Amoz, M. (1976) "A dynamic theory for composite materials," *J. Appl. Math. Phys. (CAMP)*, **27**, 83–99.

22. Truesdell, C. and Toupin, R.A. (1960) "The classic field theories," in *Encyclopedia of Physics*, Vol. III/2, Editor, W. Wliigge, Springer-Verlag, Berlin.

23. Green, A.E. and Naghdi, P.M. (1965) "A dynamical theory of interacting continua," *Int. J. Eng. Sci.*, **3**, 231–241.

24. Steel, T.R. (1967) "Applications of a theory of interacting continua," *Quart. J. Mech. & Appl. Math.*, **20**, 57–72.

25. Bedford, A. and Stern, M. (1971) "Toward a diffusing continuum theory of composite materials," *J. Appl. Mech.*, **38**, 8–14.

26. Stern, M. and Bedford, A. (1972) "Wave propagation in elastic laminates using a multi-continuum theory," *Acta Mech.*, **15**, 21–38.

27. Bedford, A. and Stern, M. (1972) "A multi-continuum theory for composite elastic materials," *Acta Mech.*, **14**, 85–102.

28. Hegemier, G.A. and Nayfeh, A.H. (1973) "A continuum theory for wave propagation in laminated composites, Case 1: Propagation normal to the laminates," *J. Appl. Mech.*, **40**, 503–510.

29. Hegemier, G.A., Gurtman, G.A., and Nayfeh, A.H. (1973) "A continuum mixture theory of wave propagation in laminated and fiber reinforced composites," *Int. J. Solids Struct.*, **9**, 395–414.

30. Ben-Amoz, M. (1970) "Stress wave propagation in uni-directionally reinforced composites," in *Recent Advances in Engineering Science*, Editor, A.C. Eringen, Vol. 5, Part II, pp. 105–128.

31. Hegemier, G.A. and Bache, T.C. (1973) "A continuum theory for wave propagation in laminated composites; Case 2: Propagation parallel to laminates," *J. Elasticity*, **3**, 125–140.

32. Hegemier, G.A. and Bache, T.C. (1974) "A general continuum theory with microstructure for wave propagation in elastic laminated composites," *J. Appl. Mech.*, **41**, 101–105.

33. Chao, T. and Lee, P.C.Y. (1975) "Discrete continuum theory for periodically layered composite materials," *J. Acoustic Soc. Amer.*, **57**, 78–88.

34. Turhan, D. (1970) *A Dynamic Theory for Directionally Reinforced Composites*, Ph.D. dissertation, Northwestern University, June.

35. Nemat-Nasser, S. (1972) "General variational methods for waves in elastic composite," *J. Elasticity*, **2**, 73–90.

36. Barker, L.M. (1971) "A model for stress wave propagation in composite materials," *J. Composite Materials*, **5**, 140–162.

37. Tsou, F.K. and Chou, P.C. (1969) "Analytical study of Hugoniot in unidirectional fiber-reinforced composites," *J. Composite Materials*, **3**, 500–514.

38. Tsou, F.K. and Chou, P.C. (1970) "The control volume approach to Hugoniot of macroscopically homogeneous composites," *J. Composite Materials*, **4**, 528–537.

39. Torvick, P.J. (1970) "Shock propagation in a composite material," *J. Composite Materials*, **4**, 296–309.

40. Munson, D.E. and Schuler, K.W. (1971) "Steady wave analysis of wave propagation in laminates and mechanical mixtures," *J. Composite Materials*, **5**, 286–304.

41. McCoy, J.J. (1973) "On the dynamic response of disordered composites," *J. Appl. Mech.*, **40**, 511–517.

42. Bose, S.K. and Mal, A.K. (1973) "Longitudinal shear waves in a fiber reinforced composite," *Int. J. Solids and Struct.*, **9**, 1075–1085.

43. Ziegler, F. (1969) "Mean waves in laminated random media," *Int. J. Solids and Struct.*, **5**, 301–302.

44. Krumhansl, J.A. (1972) "Randomness and wave propagation in inhomogeneous media," in *Dynamics of Composite Materials*, Editor, E.H. Lee, ASME, New York.

45. Brillouin, L. (1960) *Wave Propagation and Group Velocity*, Academic Press, New York.

46. Musgrave, M.J.P. (1970) "Crystal acoustics: introduction to the study of elastic waves and vibrations in crystals," Holden-Day, San Francisco, CA.

47. Malvern, L.E. (1969) *Introduction to the Mechanics of a Continuous Medium*, Prentice-Hall, Inc., Englewood Cliffs, NJ.

48. Vinson, J.R. and Sierakowski, R.L. (1986) *The Behavior of Structures Composed of Composite Materials*, Martinus Nijhoff Publishers, Dordrecht.

49. Musgrave, M.J.P. (1981) "On an elastodynamic classification of orthorhombic media," *Proc. R. Soc. Lond.*, **A374**, 401–429.

50. Ross, C.A. and Sierakowski, R.L. (1975) "Elastic wave in fiber reinforced materials," *Shock and Vibration Digest*, **7**(1), 96–107.

51. Tauchert, T.R. and Guzelsu, A.N. (1972) "An experimental study of dispersion of stress waves in a fiber reinforced composite," *Trans. ASME*, Series E, 98–102.

52. Daniel, M. and Liber, T. (1976) *Wave Propagation in Fiber Composite Laminates*, NASA CR-135086, June.

53. Takeda, N. (1980) *Experimental Studies of the Delamination Mechanisms in Impacted Fiber-Reinforced Composite Plates*, Ph.D. dissertation, University of Florida.

54. Moon, F.C. (1972) "Wave surfaces due to impact on anisotropic plates," *J. Comp. Mater.*, **6**, 62–79.

55. Mindlin, R.D. (1961) "High frequency vibrations of crystal plates," *J. Appl. Mech.*, **19**, 51.

56. Moon, F.C. (1973) *Theoretical Analysis of Impact in Composite Plates*, NASA CR-121110.

57. Chamis, C.C. (1971) *Computer Code for the Analysis of Multilayered Fiber Composites Users Manual*, NASA TN D-7013.

58. Takeda, N., Sierakowski, R.L., and Malvern, L.E. (1981) "Wave propagation experiments on ballistically impacted composite laminates," *J. Comp. Mater.*, **15**, 157–174.

59. Liu, D. and Malvern, L.E. (1987) "Matrix cracking in impacted glass/epoxy plates," *J. Comp. Mater.*, **21**, 594–609.

CHAPTER 4

DAMAGE DETECTION AND CHARACTERIZATION

4.1 INTRODUCTION

Up until now, the principal concern explored in the area of the dynamic response of composites has focused on assessing and characterizing impact damage based upon experimental/analytical models. The focus on this area is because of its importance to the performance and acceptance of composite materials for in-service use. Impact damage, albeit most important, represents but one source of damage which can occur in composites. The various damage types may be broadly classified according to Table 4.1. Individual damage sources within each category are noted below, with the observation that impact damage (highlighted) appears in each of the respective categories [1].

TABLE 4.1 Sources of Damage [1]* **(Fabrication/Processing, In-field/ Service, and Defects)**

Fabrication/Processing	Typical Defects in Composites	In-field/Service Problems
Abrasions, scratches, dents, punctures	De-bonds	Vibration
Cut fibers	De-laminations	Shock
Knots, kinks	Inclusions	Lightning damage
Improper slicing	Voids, blister	Environment cycling
Voids	Impact damage	Flight loads
Resin-rich, resin-lean areas (improper tensioning)	(tool drop, pebble impact)	Improper repair
Subquality materials	Fiber misalignment	In-storage creep or handling loads
Cure problems	Cut or broken fibers	Pebble impact (tool drop)
Inclusions, bugs, foreign contamination	Abrasions, scratches	Scratches, dents, punctures
Tool installation and removal during processing	Wrinkles	Corrosion
Mandrel removal problems	Resin cracks, crazing	Erosion, dust, sand
Machining problems	Density variations	Bacterial degradation
Shipping to propellant processing	Improper cure	
Tool drop (impact damage)	Machining problems	
Proof testing		

* Reprinted with permission from SAMPE.

4.2 DAMAGE DETECTION METHODOLOGY

Laboratory interrogation of composite failure mechanisms due to damage sources as outlined above can include, in the broadest sense, the following detection methods:

- brittle lacquer, coatings;
- strain measuring transducers;
- interferometry and holography;
- non-destructive evaluation (NDE);
- fractography.

The primary damage detection concerns are as follows:

- material characterization;

- quality control;
- component reproducibility/processing;
- damage assessment and repair;
- product acceptance;
- joints/attachments;
- environmental effect;
- non-destructive evaluation (NDE).

The use of NDE technology represents the focus of the present discussion, as a damage detection method for assessing the aforementioned areas of concern.

4.3 NON-DESTRUCTIVE EVALUATION

A general summary of the NDE techniques used for damage assessment of defects in composites is included in Table 4.2 [1]. This table is extremely useful as a screening guide to evaluate the applicability of a particular technique for use in detecting defects. The information in this table is by no means inclusive since new NDE techniques are continuously emerging. While attention in the ensuing discussion is directed toward the NDE methodology cited in Table 4.2, the importance of visual methods for surface examination should not be neglected. Visual methods can include or exclude the use of optical devices. It should also be noted that NDE techniques are used to assess both material conditions and material properties as cited in Table 4.3.

In the following sections, a description of the acoustic emission, ultrasonics and radiographic techniques is included. These descriptions include information on the NDE principle and the associated methodology.

4.3.1 Acoustic Emission

This technique is based upon the principle that in materials subjected to mechanical, thermal, or other loadings, internal changes which occur within the material emit transient sound waves.

The emitted sound waves can be detected using electromechanical transducers affixed to the surface of the materials being interrogated and the signal monitored can be electronically processed. It is impor-

TABLE 4.2 NDE Techniques for Defect (Damage) Assessment [1]*

	Radiography	Computer Topography	Ultrasonic	Acoustic Emission	Acoustic Ultrasonic	Thermograph	Optical Holography	Eddy Current
Principal characteristics detected	Differential absorption of penetrating radiation	Conventional X-ray technology with computer digital processing	Changes in acoustic Jimpedance caused by defects	Defects in part stressed generate stress waves	Used pulsed ultrasonic stress wave stimulation	mapping of temperature distribution over the test area	3D imaging of a diffusively reflecting object	Changes in electric conditions caused by material variations
Advantages	Film provides record of inspection, extensive data base	Pinpoint defect location. Image display is computer controlled	Can penetrate thick materials; can be automated	Remote and continuous surveillance	Portable quantitative, automated, graphic imaging	Rapid, remote measurement. Need not contact part, quantitative	No special surface preparation or coating required	Readily automated, moderate cost
Limitations	Expensive, depth of Jdefect not indicated, radiation Jsafety	Very expensive, thin wall structure might give problems	Water immersion of couplant needed	Requires application of stress for defect detection	Surface contact, surface geometry critical	poor resolution for thick specimens	Vibration-free environment required, heavy base needed	Limited to electric condition, materials, limited penetration depth

Defects detected							
Voids	Yes	Yes	No	Yes	Yes	Yes	Yes
Debonds	Yes	Yes	Yes	Yes	Yes	Yes	Yes
Delaminations	Yes	Yes	Yes	Yes	Yes	Yes	Yes
Impact damage	Yes	Yes	Yes	Yes	Yes	Yes	Yes
Density variations	Yes	Yes	No	Yes	No	No	Yes
Resin variations	Yes	Yes	No	Yes	No	No	Yes
Broken fibers	Yes	Yes	Yes	Yes	No	No	Yes
Fiber misalignment	Yes	Yes	Yes	Yes	No	No	Yes
Wrinkles	Yes	Yes	Yes	Yes	Yes	Yes	Yes
Resin cracks	Yes	Yes	No	Yes	Yes	Yes	Yes
Porosity	Yes	Yes	No	Yes	Yes	Yes	Yes
Cure variations	No	Yes	No	No	No	No	Yes
Inclusions	Yes	Yes	Yes	Yes	No	No	Yes

*Reprinted with permission from SAMPE.

TABLE 4.3 **Material Conditions/Material Properties Assessed by NDE Techniques**

Material Conditions	Material Properties
Anisotropy	Tensile modulus
Microstructure	Shear modulus
Grain size	Tensile strength
Porosity, voids	Shear strength
Phase composition	Yield strength
Hardening depth	Bond strength
Residual stress	Hardness
Heat treatment	Impact strength
Fatigue damage	Fracture toughness

tant to note that this method can be used to detect ongoing changes in a material; however, defects which are already present cannot be revealed. Several key issues play a role in the ability to detect the sound waves generated from internal defects including the effect of material damping and undesired noise sources such as in-service mechanical equipment and electrical line noise.

Principle of Operation In principle, acoustic emissions are transient elastic waves that arise spontaneously from localized stresses or strains. For metal-based systems, dislocation motions, micro-deformations, grain fracturing, and crack growth represent sources of acoustic emissions, while for composite materials, these sources can occur due to matrix cracking, fiber/matrix debonding, fiber fracture, fiber pull-out, and structural flaws. All of the above-mentioned sources of emission are often related to energy bursts from one or more of these sources. The frequency range of the emitted sources can span from the audible to the ultrasonic frequency range, depending on the source type and strength. The general range of detection of acoustic frequencies covers the frequency range from 20 kHz to 2 MHz. For detecting emission response, acoustic emission sensors of the piezoelectric and laser optical type are used. By introducing such probes, the relative energy variations of the acoustic emission (AE) signals occurring during material processes such as material loading and heating can be monitored and characterization of the material behavior determined.

To measure the acoustic emission energy, the method of counting the number of oscillations occurring in real time is used. The quantitative measure of acoustic activity is generally taken as proportional to the number N of oscillations per unit time. The results of the acoustic activity count are plotted with respect to time and are used to evaluate the ongoing activity of the micro-structural processes. A typical schematic of an acoustic emission system, as used for composite materials, is shown in Figure 4.1 [2].

As opposed to the activity events occurring for ongoing micro-structural processes, mechanical, thermal, and other excitation methods can be used to simulate acoustic signal generation, simulating the activity process. The principal objective is to generate acoustic waves without affecting the structural integrity of the materials as opposed to active processes which, for composites, can include matrix cracking and fiber fracture. The methodology of the simulation method is based upon variations in the oscillation counts per unit time as in the case of acoustic emissions. One such example is that of the use of a sensor, such as a quartz crystal, to investigate the activity of a unidirectional composite material on the position of the source of emission relative to that of the transducer. Both single and multiple events can be studied along with the position dependency of the transducer along the sample length. This methodology can be used to examine the correlation between AE count and the energy received by the transducer.

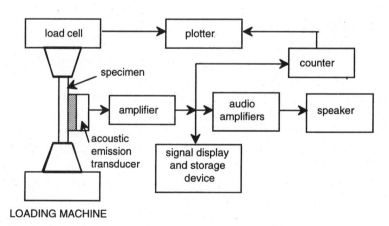

Figure 4.1 Schematic diagram of the acoustic emission (AE) detection system.

Concluding Remarks   There are a number of factors related to the interpretation of the signal patterns generated which are important in monitoring and interpreting AE signals from composite materials. Among these factors are:

1. extraneous noise within the testing equipment; *Noise*
2. emissions arising from specimen design and not failure activity within the material; *Design*
3. effects of transducer position and signal attenuation; *position*
4. wave reflections, interference, and energy losses; and *Reflector loss*
5. complex failure mechanisms which occur. *Failure*

While these factors contribute to difficulty in interpreting AE data, there are a number of areas which can be used to assess the data generated. For example, it may be possible to differentiate and relate the emissions occurring to the type of predominate damage. The failure mechanisms—due to steady state and/or varying loads—can be differentiated, as can be the relaxation and inelastic behavior of the matrix. The activity count can also be used as a proof of test technique for structural configurations.

The issues discussed above can be used to achieve transition of the AE technique from the laboratory to field conditions for material evaluations through the use of computerization to interface and speed up signal processing. Still at issue is the fact that at the present time the methodology developed within the laboratory has not been widely transitioned to commercial and industrial use.

In terms of research issues, improved theoretical models are needed in order to promote individual applications and define in service limitations. Correlations such as wave attenuation and velocity with material strength and stiffness, strength and toughness with attenuation, and other factors are needed. The requirement of increased accuracy in AE measurements requires the introduction of advanced instrumentation to account for coupling, wave transmission, bandwidth sensitivity and other factors. Also needed are standards for base line material samples. Overall, a continued focus on the importance of recognizing the basic issues of activity levels as related to failure mechanisms is needed. These and other yet unresolved issues are important to the wide introduction of AE into the industrial sector.

4.3.2 Ultrasonics

Ultrasonics is the application of sonic energy dealing with vibratory waves at frequencies higher than those audible to human beings. The frequency range is often above 16–20 kHz. Ultrasonic forms of sound signals are quite prevalent in the animal kingdom and are also used for navigation, detection and protection against dangerous situations, and for locating food. A number of studies have been done on animals, including birds, bats, moths, and dogs.

Ultrasonic application to materials and structures involving both biological and physical materials have grown almost exponentially since World War II, when the primary use of ultrasonics was to detect and locate enemy marine vessels. As a result, there exists a large library of literature [3–6] on the theory, procedures, instrumentation, and specific application techniques of ultrasonics covering a wide spectrum of intensity and frequency bands.

Basic Ideas Ultrasound waves are essentially high frequency mechanical vibrations (higher than the audible range), which can pass through different types of materials. During wave passage, the material particles oscillate due to the elastic nature of the material medium. The primary material media consisting of solids, liquids, and gases can exhibit a wide variety of wave motions. In solid materials, the particle movement can take place along lines of direction associated with the input wave as well as along lines at right angles to the direction of wave travel. The former wave is called a compressional wave or a longitudinal wave, which is the primary wave, while the latter is known as the shear, or transverse waves.

There also exist various types of surface waves that can be generated in solid materials, these being:

Rayleigh Waves   Waves which are generated at the free surface of a solid material. In a layered medium, they can be dispersive, which implies dependence of the wave velocity on the wave frequency.

Love Waves   Waves which are generated at the interface between two solid layers. The particle oscillation is confined to the plane parallel to the interface.

Lamb Waves   Waves occurring in a thin material plate (of the order of a few wave lengths). The complex wave forms containing both compressional and shear waves are the Lamb waves. These waves are also dispersive in nature.

In industrial ultrasonic applications to composite materials, the most predominate wave forms introduced are the compressional and shear waves. For anisotropic composites there exist three independent bulk waves, one being longitudinal and the remaining two being shear. Waves occurring at a material interface are becoming important in the measurement of important interface material variables, such as interfacial shear strength and debonding occurring at the fiber-matrix interface in composites.

Ultrasonic Measurements   Ultrasonic measurements can be broadly divided into two primary categories:

Material Property Measurements   Measurement devices and procedures can be designed to correlate modulated output signals with material properties; such as elastic stiffness, tensile strength, and fracture toughness. These measurements are based upon the fact that for some materials the strengths and elastic constants are well established and, therefore, by measuring longitudinal and transverse wave velocities, the strength parameters can be evaluated. These type of studies have been done for materials such as concrete, ceramics, and selected composites.

Material Damage Measurement   Ultrasonic methodology can be adapted to continuously monitor the internal changes occurring in materials. Location, size, and degree or extent of damage within the materials can be graphically mapped. Measurements can be made to estimate void content, debonding between fiber and matrix, micro-cracking, as well as type of damage.

It is important to realize that the complexity of wave motion through and around the damage zone within the material is significant, and that what is observed as output associated with the modulated signal coming out of the material is an integrated effect. A number of factors are deemed responsible for the outcoming signal moduluation and are illustrated in Figure 4.2.

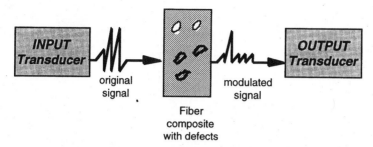

Figure 4.2 Signal modulation factors (coupling, diffraction, reflection, absorption, dispersion, scattering).

① Pulse Echo
② Through Trans mission

Methods and Instruments Among the currently prevailing methods used in the ultrasonic testing of composites, the pulse echo and through-transmission methods are most popular. These methods differ from each other basically in their modes of monitoring the modulated signal traveling through the specimen containing defects and damage. Each technique is discussed in the following sections.

Pulse-Echo Method In this method the pulsed wave generated from the ultrasonic transmitter propagates through the material specimen at ultrasonic velocity. The ultrasonic wave is partly reflected from a defect, which can be a material and/or geometric inhomogeneity, with the remainder of the wave reflected back after traveling to the boundary of the specimen. The reflected wave energy is then received at the same sending transmitters. The elapsed time as measured with reference to the echoes from defects or other interfaces, can be correlated to their locations. A schematic of a typical automatic measuring system is shown in Figure 4.3. The basic system consists of an ultrasonic transducer which acts as a transmitter as well as a receiver, a microcomputer-controlled moving fixture for scanning the entire specimen by executing motion along the *x* and *y* axes, an amplifier for amplifying the received signal, and a means for storing and displaying the signals generated from the test specimen. Water is generally used as a coupling medium between the ultrasonic wave generator and the specimen. The sensitivity of detecting flaws or damage within the materials is affected by many factors some of which are related to the shape, size, and severity of the inherent material defect or non-homogeneity. The characteristics of the received echo signal can depend on a number of factors, some of which are listed below [5]:

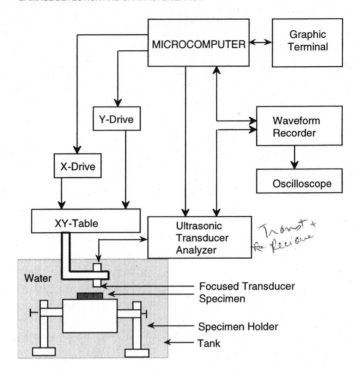

Figure 4.3 Schemamtic diagram of a pulse–echo automated measuring system.

- power of the transmitted pulse into the specimen;
- directional feature of the transmitter probe;
- size, shape, and orientation of the inhomogeneity;
- losses in echo signal at the receiving transducer due to signal directionality, reflection, and coupling;
- wave attenuation attributed to scattering, absorption, and reflection inherent to heterogeneity of composite material specimen.

Through-Transmission Methods   This method requires the use of two ultrasonic transducers such that one of them acts as a transmitter and the other as a receiver. The relative positioning of these transducers with reference to a specimen depends upon whether the longitudinal, shear, or surface waves are being used in the interrogation of the specimen. The transmitter and receiver transducers can be located on opposite faces of the specimen for measurements taken in a longitudinal wave mode, or they may be located on the same face for measurements in a shear wave mode. For measurements in a surface wave mode, these transducers can be located on the same face of the specimen but at the opposite ends of the specimen. A schematic for an automated scanning system using ultrasonic transducers is shown in Figure 4.4. The primary elements of this measurement system consist of essentially the same features as used in the pulse-echo system except for the introduction of two transducers. The material medium necessary to couple the ultrasonic wave energy with the specimen as mentioned, is generally water. For automated scanning of the entire test area of the specimen microcomputer-controlled (x–y) drives are used.

All the elements mentioned in an ultrasonic system, including pulse generators, amplifiers, microcomputer, data storage, and recording equipment and transducers need special attention in selection such as

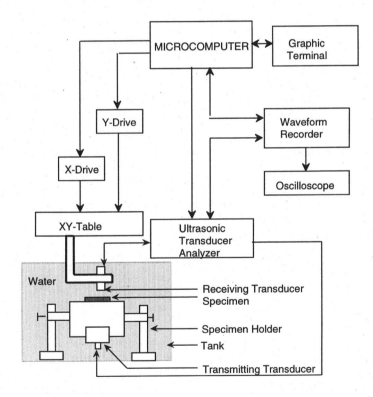

Figure 4.4 Schematic diagram of a through-transmission automated measuring system.

① Image
② Time
③ Freq

① ID A
② T DA
③ FDA

their ranges, sensitivities, and speeds etc. before a good scan of the hidden defects and damage in composites can be obtained.

The overall ultrasonic data processing, and analyses can normally be done by using three basic approaches: image-domain analysis, time-domain analysis, and frequency-domain analysis. The effective uses of these techniques for material and defect or damage characterization can be illustrated through Figure 4.5 [7].

Ultrasonic measurements for evaluating defects and damage in fiber-reinforced composite materials present additional challenges as related to requirements associated with their sensitivity and resolution. These problems are primarily due to the inhomogeneity of the composite material properties, variations in the degree of cure, variations in the fiber–matrix bond strength, and other defects which can occur during the manufacturing process. An enormous amount of literature dealing with mechanical property measurements, property–defect correlations, location and characterization of defect and damage in a wide variety of polymer composites exists, with applications of this technology to the aerospace and missile industries currently routine. With respect to the ultrasonic methods discussed, the pulse-echo methods are limited to application involving thin composites due to the poor resolution of waves penetrating to greater depths. In such cases, through-transmission techniques have an advantage. References such as Ensminger [4], Krautkramer and

Thin plates = Through Transmission Method is Old

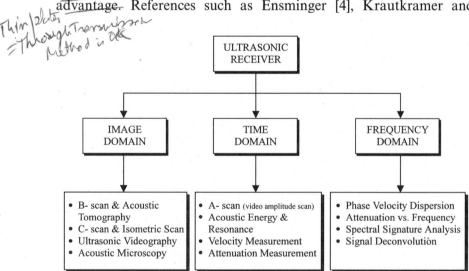

Figure 4.5 Ultrasonic data processing and analysis methods [7]. (Reprinted with permission from Plenum Press.)

Krautkramer [5] and the references listed in these publications cover a wide variety of applications of this technology as applied to composite materials/structures.

4.3.3 Radiographic Techniques

X-rays and gamma-rays are commonly used to perform non-destructive evaluations on the quality and integrity of a large class of materials, including fiber-reinforced composites. X-ray electromagnetic beams, with a wide range of wavelengths, can penetrate through composites and produce photographic images of the internal structure of the material specimen. This image, made on a photographic film or other types of electromagnetic imaging media, can reveal the location and size of a defect or existing damage. Material objects are also known to absorb these radiations to varying degrees, which depend upon the wavelength of the radiation, the material composition, and the thickness of the targeted specimen.

The waveband of x-rays used can be very broad, ranging from about 10^{-4} nm ("soft" x-rays) to about 10 nm ("hard" x-rays) [6]. The "hard" or high-energy x-rays are used for very thick material specimens. The radiation beam impinging upon a specimen goes through absorption, scattering, and transmission, which is captured on film. The image found on the film can then be interpreted.

The number of x-rays absorbed and transmitted by a material depends upon its atomic number. Therefore, to make the damage—such as matrix debonding, fiber breakage, voids, flaws or delamination in fiber-reinforced polymer composites—more visible, an x-ray enhancing penetrant is injected into the specimen before the radiograph is taken. These high atomic number compounds improve the dark and gray contrast on the resultant radiograph by creating differential absorption bands within the material. The preparation of the specimen for such an interrogation consists primarily of the following four steps:

- preparation of specimen by removing moisture and surface contamination;
- soaking the specimen with the penetrant solution for a given time;
- x-raying the specimen;
- removing the impregnant from the specimen.

The resultant radiographs show dark lines and/or regions within the gray image of the specimen, which identify the regions of compound penetration. The real task of getting good radiographs for a given composite specimen involves to some degree of trial and error until the correct combination of currents, voltages, film-to-focus distances, and exposure time has been reached. These variables have to be readjusted if the composite type and/or the thickness of the specimen are changed. Readers requiring a more comprehensive background on the theory and practice of general radiographic techniques can consult any number of monographs on the subject, such as the one by Halmshaw [6].

The issue of radiation safety is an important issue with the operating use of such devices. Guidelines, recommendations, and general safety regulations provided by various local, national, and international agencies need to be rigorously followed.

4.4 DAMAGE INITIATION AND GROWTH MECHANISMS

The low-velocity, high-velocity, and hyper-velocity impact loading regimes are known to be marked by their distinctive signature features of damage and failure modes in fiber-reinforced composites. These distinctive features include the damage/failure modes in both local and global domains, which are known to depend upon many geometrical and material parameters such as matrix properties, fiber and fiber-matrix interface properties, stacking sequence, support conditions, and rigidity, shape and size of the impactor. It is therefore of great importance to identify the key material, geometrical, and impactor-target interaction parameters that can give us a basis to look at the concept of "designing composites with a predetermined impact tolerance" for various impact loading situations.

Although considerable research work has been reported [8–12] on the effects of fibers, resins, fiber–matrix interface, stacking sequence, and hybridization on the damage and failure modes, the fundamental parameters for designing against a predetermined impact tolerance have yet to be identified. One reason for the lack of this understanding is the fact that different investigators have used widely different impactor-target systems, which does not provide the kind of coherence necessary to delineate relevant control parameters. However, some key observations on the type of damage initiation and the growth mechanisms occurring in various composite systems under a given impact loading situation can be delineated. As an illustration, the investigation discussed herein reflects the following salient features related to a controlled impact study:

- The impact experiments have been designed to observe and document damage initiation and growth mechanisms in glass-, graphite-, and Kevlar-epoxy composite plates, subject to the same loading conditions.
- The stacking sequences have been chosen to vary the number of interfaces for a fixed total thickness of cross-ply laminates for all three materials. This requirement has resulted in some variation in the total number of plies in these material systems because the ply thickness for the glass-epoxy, graphite-epoxy, and Kevlar-epoxy prepreg tapes was not the same. A quasi-isotropic angle-ply laminate configuration for glass-epoxy was also added to the study so that the shape and size of the damage zone could be compared with the corresponding cross-ply laminate configuration.

4.4.1 Materials and Laminate Configurations

The prepreg tapes chosen in this study contained the following types of fibers and resins (Table 4.4). The prepreg tapes were cut to plate size (30 × 30 cm) and stacked to form laminates, which were autoclave cured. A pressure and temperature cycle recommended for each of the composite systems studied was followed in fabricating the laminated plates. Specimens of size (15 × 15 cm) were cut from these plates for impact studies with the stacking sequences tested listed in Table 4.5. These stacking sequences were designed to control the number of interfaces. Therefore, the variation in the total number of plies for these material systems was controlled to keep the total

TABLE 4.4 Fiber and Matrix Materials

Material	Fiber	Matrix
Graphite-epoxy	THORNEL 300	PR-322 Epoxy
Kevlar-epoxy	KEVLAR 49	PR-328 Epoxy
Glass-epoxy	E-GLASS	Type 1003

TABLE 4.5   Laminate Configurations for Composite Plates

Material	Stacking Sequence
Glass-epoxy	$[0/90/0—]_{15}$, alternating ply
	$[0_3/90_3/0_3/90_3/0_3]$, four interfaces
	$[0_5/90_5/0_5]$, two interfaces
	$[0/145/90]_{2s}$, quasi-isotropic
Graphite-epoxy	$[0/90/0—]_{27}$, alternating ply
	$[0_9/90_9/0_9]$, two interfaces
Kevlar-epoxy	$[0/90/0—]_{17}$, alternating ply
	$[0_4/90_4/0_4/90_4/0_4]$, four interfaces
	$[0_6/90_6/0_6]$, two interfaces

thickness of each laminate about the same (thickness 0.132–0.152 in) and thus, the effect of thickness variation on the bending stiffness minimized.

4.4.2   Introduction and Observations of Damage

In this study attention has been focused on localized impact loading situations in the sub-perforation impact velocity range. A plane-ended steel cylindrical impactor with a length of 25.4 mm and a diameter of 9.525 mm was selected to produce impact damage at the center of 15 × 15 cm composite plates. The impactors were propelled through a gas gun barrel with the gas gun chamber pressure adjusted to obtain the desired range of impact velocities. Each composite plate specimen was impacted at three pre-selected impact velocities: 45.73 m/s (150 ft/s), 68.6 m/s (225 ft/s), and 91.46 m/s (300 ft/s), respectively. The impact velocity measurements were made using a sensing device, which consists of photocells used in conjunction with light beams separated at a fixed distance. Details concerning the gas gun assembly and the plate holding fixture can be found elsewhere [13]. It should, however, be mentioned that all four sides of the target plate specimens were clamped into a steel frame specimen holder, thereby leaving the effective specimen size as 14 × 14 cm.

As discussed earlier several non-destructive evaluation techniques, such as ultrasonics, acoustic emission, and x-ray radiography, are available for use in obtaining damage-zone shape and size in composite materials. The suitability for use of a particular technique depends upon many factors, such as the kind, degree, and location of damage through the thickness of the laminates. Based upon recent advances and experience with a wide variety of composite systems, it is generally recognized that, due to the presence of several damage modes, only a suitable combination of NDE techniques can provide, at best, the necessary information on the damage characteristics. The focus in this experimental investigation has been directed toward obtaining a qualitative picture of the dominant damage modes related to the three primary composite systems catalogued.

The semitransparency of the glass-epoxy and Kevlar-epoxy composites was used to obtain information on the type of delamination and damage spread with respect to shape and size. A high-intensity uniformly illuminated light source was used to obtain an appropriate photographic image of each of the damaged specimens. The light intensity and the film exposure were adjusted to obtain qualitative information on the delamination growth at different interlaminar planes within the impacted plates.

4.4.3   Cylindrical Bend Tests

The cylindrical bend tests were performed on all impacted specimens in order to make observations on further growth of their existing damage modes. The test specimens were supported along two edges, selected in such a way that the plate was tested in the stronger fiber orientation. The bending test fixture was mounted in the testing frame of a testing machine and the load recorded directly from the load cell of the machine, while the mid-point displacement of the specimen was recorded by a dial gage. The "bending strength" of the plate was taken as the maximum load P_{max}, and the residual strength factor was taken as P_{max}/P^*_{max}, where P^*_{max} was obtained for the undamaged plate.

4.5   COMPARISON OF DAMAGE MECHANISMS AND THEIR GROWTH IN GLASS-, GRAPHITE-, AND KEVLAR-EPOXY MATERIAL SYSTEMS

Typical photographs obtained showing damage shapes and sizes for the glass-epoxy and Kevlar-epoxy plates at three different impact velocities are shown in Figures 4.6(a,b), 4.7, and 4.8. The following observations can be made:

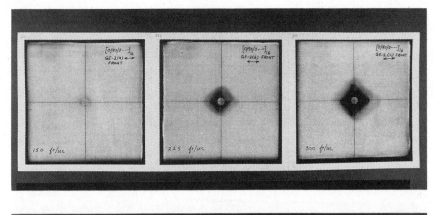

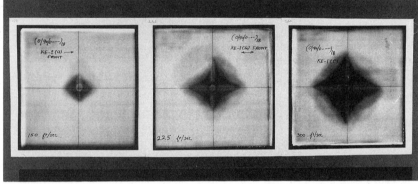

Figure 4.6 Impact damage in glass-epoxy (top) and Kevlar-epoxy (bottom) cross-ply laminated composite plates (plane-ended cylindrical steel impactor with 9.525 mm diameter and 25.4 mm length).

The size of the delamination damage zone expands with increase in the impact velocity for the material systems studied, however, the shapes for a given stacking sequence do not change.

The shape of the delamination damage depends upon the lamination sequence, and also on the change in the number of interfaces. For the glass, as well as the Kevlar-epoxy systems with the number interfaces 15 and 17, respectively (Figure 4.6a, b), equiaxed rectangular shapes occur, which is in contrast with the more elongated shapes when only four interfaces are present (Figure 4.7). In addition, the delamination size is found to increase in each successive interlaminar plane through the thickness as its location regresses relative to the impacted front surface. The last interlaminar plane, at the back of the impacted

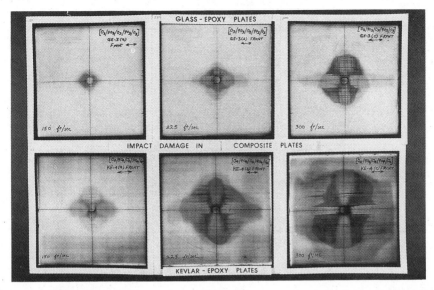

Figure 4.7 Comparative illustration of impact damage in cross-ply (4 interfaces) laminated glass- and Kevlar-epoxy composite plates (impact velocities 150, 225m and 300 ft/s, respectively).

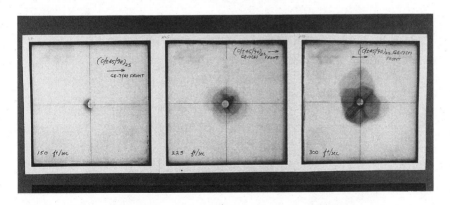

Figure 4.8 Impact damage in quasi-isotropic $(0/\pm 45/90)_{2s}$ glass-epoxy composite plates.

specimen suffers the largest delamination. This type of delamination process involves the formation of a generator strip with sequential delamination occurring. This process is described in brief in Section 5.4 (Figure 5.25) and in detail by Cristescu et al. [13].

The Kevlar-epoxy material system for all stacking sequences listed and impact velocities exhibits a much larger delamination type damage when compared to the corresponding glass-epoxy system. This appears to indicate that the former material system has a relatively weaker interlaminar bond strength.

- A comparison with a graphite-epoxy system with a cross-ply, three-interface configuration shows the existence of a shear cut-out, a plug under the impact point (Figure 4.9). Indeed, the blunt-ended projectile is shown to have penetrated through the plate and delaminated a whole strip, from the back face of the plate. This type of "punching a whole strip though the plate" was not observed for either the glass-epoxy or Kevlar-epoxy plates for the impact velocity range tested.

- Post-impact cylindrical bend tests performed on all the impacted material systems tested exhibited growth in the existing delami-

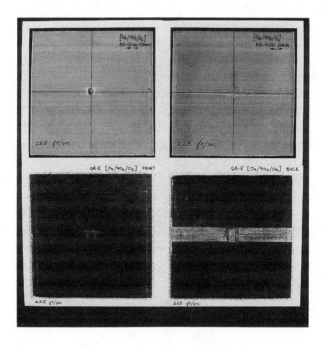

Figure 4.9 Comparative illustration of impact damage in cross-ply (two interfaces) Kevlar- and graphite-epoxy composite plates (impact velocity 225 ft/s, top photographs: Kevlar-epoxy, bottom photographs: graphite-epoxy).

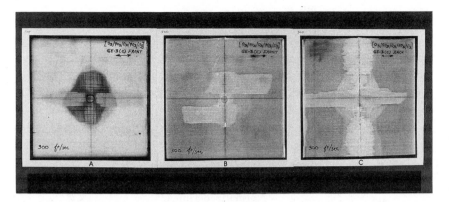

Figure 4.10 Delamination growth and fiber fracture in cross-ply (four interfaces) glass-epoxy composite plates under post-impact cyclindrical bend tests (A: damage zone after impact; B: front face or loading face; C: back face; impact velocity: 300 ft/s).

nation during testing with accompanying fiber fracture on the tension face of the plate. Typical failure modes for a glass-epoxy specimen have been shown in Figure 4.10.

4.6   IDENTIFICATION OF IMPACT DAMAGE MODES AND MATERIAL PARAMETERS

The following significant conclusions about the damage modes and material parameters governing these observed failures can be cited.

4.6.1   Damage Modes

The most dominant damage modes observed were

- Local damage modes
  Generator strip formation
  Fiber fracture under the impact zone (partial penetration)
- Global damage modes
  Matrix cracking (clearly visible on the front and back surfaces of the plate)
  Delamination in the interlaminar planes
  Bulging of the entire plate specimen

- Damage modes under post-impact flexural loading
  Delamination crack-growth
    stable and
    unstable
  Fiber fracture
    stable and
    unstable
  Large permanent out-of-plane deformation

4.6.2 Important Material Parameters

- For local damage
  Fiber fracture strain
  Transverse tensile strength of laminate
- For global damage
  Interlaminar bond strength
  Delamination crack-growth resistance

4.6.3 Comparative Assessment

- For all three material systems glass-, graphite-, and Kevlar-epoxy, the strength degradation due to impact damage is greater than stiffness degradation.
- The residual strength factors P_{max}/P^*_{max} (defined earlier) are found to depend upon the stacking sequence, composite type, and impact velocity.
- Generator strip formation is found to have a correlation with the fracture strain of the fibers. This strip is more pronounced with the material system having greater fiber fracture strain for a given fiber-matrix bond strength. With reference to the generator-strip formation, the three material systems ranked as follows: glass-epoxy > Kevlar-epoxy > graphite-epoxy.
- The material systems with reference to out-of-plane large deformation (or bulging) can be ranked as: Kevlar-epoxy > glass-epoxy > graphite-epoxy.
- For matrix crank density, the glass-epoxy showed a higher density compared with the Kevlar-epoxy system.

TABLE 4.6 Failure Modes During Flexural Loading

Material	Delamination-Crank Growth (Stable and Unstable)	Fiber Fracture (Stable and Unstable)
Glass-epoxy	Yes	Yes
Kevlar-epoxy	Yes	No
Graphite-epoxy	Yes	Yes

- Perforation resistance of the materials could be ranked as: glass-epoxy > Kevlar-epoxy > graphite-epoxy.
- With respect to delamination size, the materials could be ranked as: Kevlar-epoxy > glass-epoxy.

The failure modes observed during flexural loading of already impact-damaged specimens can be summarized as in Table 4.6.

4.7 CONCLUDING REMARKS

Although much work has been reported on the effects of fibers, resins, stacking sequence, and hybridization on materials response, a full understanding about the controlling parameters for designing against impact loading has yet to emerge. One of the primary reasons for lack of such information is the fact that different investigators have used widely different impactor-target systems; hence, a systematic and unified understanding has not emerged. However, some interesting observations should be noted; for example, in graphite, glass and Kevlar composite systems, the delamination type of failure mode appears to be dominant in the sub-perforation range. In addition, both glass and Kevlar systems appear to have better impact resistance than graphite, although the latter is stronger and stiffer. Any approach to systematize this behavior should include the fact that graphite fibers have relatively low fracture strains and are thus more prone to a localized failure mode.

The damping characteristics of individual systems could also be of significant importance in evaluating the impact response. Hybrid systems of glass, graphite, and Kevlar fibers should be more efficient when stiffer and stronger graphite lamina can be sandwiched between glass and Kevlar lamina. This would result in a composite system with higher stiffness and strength as well as higher damage tolerance. In cases where delamination types of failure modes are to be avoided, or minimized, introduction of three-dimensional weaves should be advantageous.

The assessment of damage and the characterization of residual mechanical properties requires a careful screening of the many NDT interrogation techniques currently available. In this regard an attempt by Hamstad et al. [14] to identify some correlations between the residual strength of impact-damaged Kevlar-epoxy composites and acoustic emission (AE) events is worth noting. These conclusions affirm that there exists various degrees of correlation between AE and residual strength at different states of damage, and these can be identified through monitoring of average event rates, total duration events, and event rate moments. Caprino and Teti [15] made further attempts to correlate AE activity with the residual tensile strength of impact-damaged GRP laminates. Their experimental results show that a stronger correlation exists between the AE activity and residual tensile strength after impact than that with impact energy. Other interesting developments for NDE applications include monitoring the processing of composite materials. Exploratory studies by Canumalla et al. [16], conducted on metal-matrix composites, indicate a strong potential for the development of an in situ AE-based technique for the infiltration process. Further work on metal-matrix composites by Krishnamurthy et al. [17] indicates the usefulness of ultrasonics as a quick and reliable tool for monitoring consolidation of these composites. It is generally recognized that no single scanning technique is ideal for composites because of the complexity and the interactive nature of the failure modes. A combination of such techniques should be identified for use with the various composite material systems in use. Of these techniques, the ultrasonic and radiographic techniques appear to be the most acceptable. Also, added emphasis needs to be placed upon further development of these techniques so that a layer-by-layer (through-the-thickness) interrogation of damage growth in laminated composites can be accomplished.

In summary, it can be stated that the goal of achieving a comprehensive understanding of the effects of various parameters including impactor, target, and impactor–target interaction on the damage tolerance of composites is still evolving.

4.8 REFERENCES

1. Munjal, A.K. (1986) "Use of fiber reinforced composites in rocket motor industry," *SAMPE Quarterly*, **17**(2), 1–11.
2. Fuwa, M., Bunsell, A.R., and Harris, B. (1976) "An evaluation of acoustic emission techniques applied to carbon-fibre composites," *J. Appl. Phys.*, **9**, 353–364.
3. Boving, K.G. (editor) (1989) *NDE Handbook*, Butterworths, London.
4. Ensminger, D. (1988) *Ultrasonics* (2nd edn), Marcel Dekker, Inc., New York.
5. Krautkramer, J., and Krautkramer, H. (1990) *Ultrasonic Testing of Materials* (4th edn), Springer-Verlag, Berlin.
6. Halmshaw, R. (1991) *Non Destructive Testing* (2nd edn), Edward Arnold, London.
7. Vary, Alex (1980) "Concepts and techniques for ultrasonic evaluation of material mechanical properties," in *Mechanics of Nondestructive Testing*, Editor, W.W. Stinchcomb, Plenum Press, NY, pp. 123–141.
8. Sun, C.T. and Sierakowski, R.L. (1975) "Recent advances in developing FOD resistant composite materials," *Shock Vibration Digest*, **7**(2), 1–8.
9. Takeda, N. and Sierakowski, R.L. (1980) "Localized impact problems of composite laminates," *Shock Vibration Digest*, **12**(8), 3–10.
10. Sierakowski, R.L. and Chaturvedi, S.K. (1983) "Impact loading in filamentary structural composites," *Shock Vibration Digest*, **15**(10), 13–31.
11. Cantwell, W.J. and Morton, J. (1991) "The impact resistance of composite materials: a review," *Composites*, **22**(5), 347–362.
12. Abrate, S. (1991) "Impact of laminated composite materials," *Appl. Mech. Rev.*, **44**(4), 155–190.
13. Cristescu, N., Malvern, L.E., and Sierakowski, R.L. (1973) "Failure mechanisms in composite plates impacted by blunt-ended penetrators," *Foreign Object Impact Damage to Composites*, ASTM STP-568, pp. 159–171.
14. Hamstad, M.A., Whittaker, J.W., and Brosey, W.D. (1992) "Correlation of Residual Strength with Acoustic Emission from Impact-Damaged Composite Structures under Constant Biaxial Load," *J. Comp. Mater.*, **26**, 2307–2328.
15. Caprino, G. and Teti, R. (1995) "Residual Strength Evaluation of Impacted GRP Laminates with Acoustic Emission Monitoring," *Composites Science and Technology*, **53**, 13–18.
16. Canumalla, Sridhar, Pangborn, R.N., Tittmann, B.R., and Conway, Jr, J.C. (1994) "Acoustic Emission for In Situ Monitoring in Metal-Matrix

Composite Processing," *Composites Science and Technology*, **52**, 607–614.

17. Krishnamurthy, S., Matikas, T.E., Karpur, P., and Miracle, D.B. (1995) "Ultrasonic Evaluation of the Processing of Fiber-Reinforced Metal-Matrix Composites," *Composites Science and Technology*, **54**, 161–168.

CHAPTER 5

DAMAGE TOLERANCE

5.1 INTRODUCTION

The concept of damage tolerance can be called multifaceted in that it carries a number of defining or characterizing features, and hence a single unifying idea for its definition has not emerged as yet. The difficulties in assigning a single all-encompassing definition lie in the fact that a multiplicity of damage modes are observed to occur within a laminated composite system, and these damage modes depend upon a number of material and design parameters that include manufacturing, assembly, loading and end-use field conditions. The extent and type of damage which can be tolerated within a structural component is not simplistic. Recognizing these difficulties, the current chapter focuses on the ideas, approaches, and models related to various aspects of the damage tolerance concept with a view to identifying essential features in order to provide a basis for a damage tolerance concept.

The concept of damage tolerance in fiber-reinforced composites has been driven primarily by the aerospace industry, and encompasses a design philosophy based on ensuring safety and integrity of a material structural system. Important to the concept of damage tolerance is the presence and/or introduction of damage into a structural system. Such sources of damage commonly encountered can be classified as those occurring and due to:

- fabrication/processing;
- assembly;
- normal service/field conditions.

Some of the typical defects and damage modes encountered within the framework of these damage sources include

- fiber misalignment;
- improper cure;
- density variations;
- voids, blisters;
- inclusions;
- debonds;
- resin cracks, crazing;
- delaminations;
- cut or broken fibers.

Of the above sources of damage and associated defects and damage modes encountered, impact damage is generally recognized to be the most severe loading condition for composite structures. The resultant damage modes that occur due to impact events include resin cracking, fiber-matrix debonding, delamination, and fractured fibers.

The defects and damage modes cited above represent the important elements necessary to document the damage state of a structural system. However, characterization of the damage state, by both visual and non-visual non-destructive test methodologies is equally important. Particularly relevant to the complex fracture processes occurring in composites are such damage detection methodologies as

- mechanical;
- optical;
- acoustical;
- ultrasonic;

- radiographic;
- thermographic.

A number of these methodologies are described elsewhere. However, to complete the quantification of damage it is not only recognition and detection of the degree of damage but the assessment of the damage state through material/structural characterization which is essential. Among the methods mostly used by the composite community to assess the damage state are tests which can evaluate basic material/structural properties. Such tests include appropriate coupon test specimens as well as structural elements which can be used to evaluate certain residual characteristics of the material system as well as the structural elements. These characteristics include principally the following:

- residual stiffness;
- residual strength/failure;
- residual toughness;
- residual dynamic response.

Included in these measurements are basic properties tests such as

- tension;
- compression;
- flexure;
- shear.

Specifications related to the damage-tolerance issue for composites is currently an evolving event. Since the major application of advanced composites has been in aircraft components, it is recognized that specific requirements on damage-tolerance specifications have been introduced in the aerospace area. In particular, the basis for composite design specifications has evolved from those established for metallic structures and cited in MIL-A-83444. Also, MIL-STD-1530A is often used for its basic philosophy and general materials requirements. These specifications focus attention on the problem of cracks in metals; however, this issue is not the critical damage tolerance measure in laminated composites. For composites, it has been observed that service-related impacts play a major role in producing defects. Such defects may not be visually detectable on the surface of the structure but can cause considerable internal damage. This latter event can lead to potential failures at loads considerably less than the structural undamaged strength. All defect types cited create problems; however, impact induced events which produce matrix cracking, delamination, and fiber fracture remain the burning issue in damage tolerant design.

One damage tolerance approach introduced by the aerospace industry is to establish two quantifiable measures as threshold levels for an assumed non-detectable damage event, these being

- a prescribed impact energy level;
- a visual damage level.

As damage thresholds, a damage level of 0.10 in indentation is used for the visible damage threshold while a 100 ft-lb energy level is considered for the energy threshold. The thresholds selected represent damage levels which ensure a safety level equivalent to or greater than that for metal structures. This is demonstrated graphically in Figure 5.1 in which a schematic of the visual damage versus impact energy, with composite thickness as a parameter.

With reference to Figure 5.1 the indentation depth as a fixed number cannot be uniquely specified for a given impact energy level because of variations in composite thickness and corresponding composite stiffness. For example, using the schematic figure, the visible damage level X for indentation would require the impact energy levels E_1, E_2, E_3, and E_4 respectively as one tends towards relatively

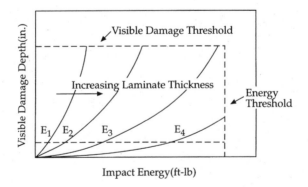

Figure 5.1 Schematici of visible damage vs. impact energy.

thicker composites. This indicates that any search for a unique damage-tolerance identification should include other parameters such as thickness in addition to indentation depth and energy levels as criteria. Presently, such criteria are being evolved not only in the aerospace area, but within the composite community at large. These criteria can vary and are application dependent.

5.2   DAMAGE-TOLERANT DESIGN

The concept of damage-tolerant design is based upon an initial assumption that damage exists and that structural design is modeled on the basis of the presence of an assumed damage level. Damage size and location as issues are specified by the authority, user or owner of the structural system. As an example, the size selected to represent damage can be arbitrary when the location is fixed such as to focus on a critical area. The location site selected to represent a critical damage state may not always be in the most logical and representative section.

Of the typical damage sources the in-service impact damage source represents a severe if not dominant design concern. A quantitative assessment of the damage tolerance issue for this damage source encompasses the following types of characterization:

- damage size and shape
  local
  global
- damage type
  matrix cracking
  delamination
  fiber fracture
- residual strength/stiffness
  material
  structural
- residual toughness
  mode I
  mode II
  mode III

combined modes

delamination fracture in Mode I, II, III and combined modes

flaw sensitivity

residual dynamic response

vibrations

wave propagation

repeated impact

One can see from the categories listed above that the damage-tolerance issue involves measurements of key residual material as well as structural characteristics. The material versus structural aspects are considered interactive, and hence may not be clearly distinguished under many practical situations. Material degradation does however, affect the structural performance; whether such an effect is significant or not depends on many physical and loading factors within the design limits. Currently, residual characteristics appear to be important for arriving at appropriate criteria for a damage-tolerant design, and are discussed.

5.3   MEASURES TO ASSESS DAMAGE TOLERANCE

For quantitative evaluation of damage tolerance of composite structural elements, identification of key parameters becomes essential. These measures can be multiple in nature, and their appropriateness to practical situations can be both material- and application-specific. For example, assume that a structural composite will be subjected to a dominantly compressive loading during its service life; the question is what characteristics represent the appropriate measure or measures that can be used to assess the damage tolerance of such a component. The degradation of the compressive stiffness and strength due to impact damage and their effects on structural performance such as buckling can be considered to be one of the appropriate measures, provided no further damage growth takes place during the remaining service life of that component. This is an example of a measure that one can classify as being a load- or application-driven measure. It should be recognized that for a given structural component introduction of damage may affect all the material and structural characteristics to different degrees, and the issues related to severity or criticality of these effects may need additional considerations such as safety and reliability other than those directly related to change in material and structural characteristics. With such interactive elements present in the damage-tolerance concept, it is not surprising that a unifying singular damage-tolerance criterion has not yet emerged. It would appear that a multifaceted approach to damage tolerance may be the most appropriate approach. A number of such measures appear appropriate, and each is discussed in the following sections:

- damage size
- residual strength/stiffness
  compression
  tension
  bending
  shear
- residual dynamic response
  vibrations
  wave propagation

5.3.1   Damage Size

One measure used to assess damage tolerance is related to the size of the damaged area in a composite component. The damage states mentioned earlier can be illustrated through Figure 5.2 where the

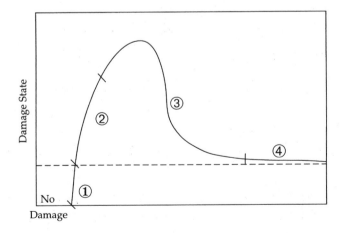

Figure 5.2   Schematic of change of damage state with impact energy.

influence of energy levels on the type of damage are plotted. These damage types, corresponding to states 1, 2, 3, and 4, are shown in Figure 5.3. The damage states identified in Figure 5.2 relate to both visible and non-visible defects. Specifically, at very low impact velocities, no damage is visible, however, as the velocity increases matrix cracking occurs. As the impact velocity is further increased, the density and extent of the matrix cracks may progress such that interfacial debonding can occur. This, in turn, leads to delamination, then fiber breakage, and finally, penetration/perforation of the target material. Thus, the damage as measured by some geometric measure, first increases and progresses to a maximum damage size and then collapses to a damage state dominated by perforation of the material/ structural elements. This latter damage state can approach asymptotically that obtained for a hole drilled in a material/structural member. It is important to note that the damage size and damage type observed in impact events are influenced by a number of factors associated with both the impactor and target. For example, the constituent materials, geometrical arrangement of fibers, and the fiber spacing can be important damage tolerant factors associated with the target. For example, a typical data sheet with typical information included for studying the damage development and characterization for composite targets subjected to foreign object damage (FOD) type of loading situations should include the following information:

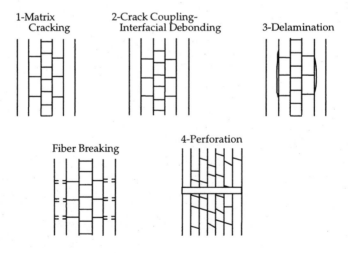

Figure 5.3 Schematic of various damage states.

- impactor properties – (e.g. steel, 3/8 in dia., 1 ft 2 in long)
- impact location – (e.g. center of plate)
- impact orientation – (e.g. non-oblique)
- load classification (impact)
- target (composite) properties – (e.g. E glass-epoxy)
- ply configuration – (e.g. cross ply)
- stacking seqeunce – (e.g. $(0_5/90_5/0_5)$)
- target thickness – (e.g. 0.135 in)
- fiber content – (e.g. 55% vol.)
- constituent properties (e.g. E glass, $E_f = 10 \times 10^6$ lbf/in^2, epoxy, $E_m = 0.5 \times 10^6$ lbf/in^2)

As a further example, consider the case of transverse plate impact for uniformly spaced fibers, arranged in a cross-ply stacking sequence, first with extensible steel fibers and then with inextensible glass fibers in an epoxy matrix. For the case of the extensible fibers, a symmetrical damage pattern occurs without significant delamination. On the other hand, for the case of the inextensible fibers, the damage appears asymmetrical with significant delamination [1]. For the latter system, delamination areas as detected by c-scan and/or radiographs can be related to the initial kinetic energy of the impactor. Some typical data obtained from Liu [2] relating these effects for the case of thermoset matrix composites are shown in Figure 5.4. Data obtained for other thermoset matrix composites as well as thermoplastic composites are shown in Figures 5.5 [3] and 5.6 [4]. For the systems studied and displayed in Figure 5.6, and for composite plates with similar stacking sequences and thicknesses, the differences in delamination are related to the material properties or mismatch of material properties between the principal fiber directions and the transverse fiber directions in the composite system. Thus, for composites subjected to a given impact energy with assumed equivalence in bond strength, the resultant delamination area would be expected to follow the general trends in mismatch between the principal and transverse composite properties. The results for a wide variety of materials (Table 5.1) are shown in Figure 5.6. Discrepancies in results noted in Figure 5.6 can be attributed to differences in the bond strength between the fibers and the matrix. For such systems, there appears to be a threshold level at which delamination initiates [5]. Also, for thermoset matrix composites, a linear variation in delami-

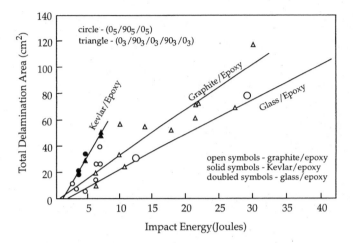

Figure 5.4 Delamination area vs. impact eneregy for thermoset composites [2]. (Reprinted with permission from Technomic Publishing Co., Inc.)

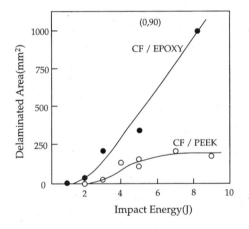

Figure 5.5 Effect of composite type (thermoset or thermoplastic) on delamination [3]. (Reprinted with permission from Elsevier Science Ltd, Kidlington OX5 1GB, UK.)

nation area with respect to impact energy has been observed. A relation of the type $(K = g_1 + g_2 A)$ can be assumed where g_1 is the threshold kinetic energy required to introduce a detectable damage while A is the total delaminated area and g_2 an apparent surface fracture energy [1]. The above relation may not be true for the case of thermoplastic matrix composites as shown in Figure 5.5. The following observations appear significant.

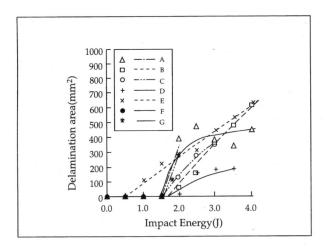

Figure 5.6   Delamination area vs. impact energy for graphite-composites [4]. (Reproduced with permission of the Royal Aeronautical Society.)

- A threshold level for delamination damage initiation is observed for thermoplastic as well as thermoset composites.
- The delamination size increases with increase in impact energy for all the stacking sequences and for both the thermoset and thermoplastic composites. Delamination size, however, is greatly reduced for a cross-ply thermoplastic composite (CF-PEEK) as shown in Figure 5.5. The parameters associated with this reduction can include the increased inter-laminar shear strength and toughness of thermoplastic resins.
- The effects of fiber types and stacking sequence as a function of the delamination size are shown in Figure 5.4. It is observed that a Kevlar-epoxy composite exhibits a relatively larger delamination area when compared with the corresponding glass and graphite-epoxy systems. This difference can be attributed to the relatively poor interlaminar strength of Kevlar composites. The effect of stacking sequence on the total delamination size, as reflected through the number of delaminated interlaminar planes, is not apparent.

It has been shown that when pre-loads are introduced into the specimen, the overall threshold level for damage initiation is lowered [6]. In order to obtain data for pre-loaded specimens, a large number of

TABLE 5.1 Summary of Composite Test Panels [4][*]

Lay-up Code	Orientation	Number of Lamina	Thickness	Fiber	Resin Matrix
A	$(\pm45/0_3)_s$	10	1.55	Thornell 300 Tape all unidirectional	Hexcell F263
B	$(\pm45/0_3)_s$	8	1.71	Thornell 300 U/D Tape/8 harness Satin weave	Hexcell F263
C	$((\pm45/0)/(090)/(\pm45/0))$	4	1.54	Thornell 300 U/D Tape/8 harness Satin weave	Hexcell F263
D	$((0,90_3{}^k))_s$	6	1.65	Kevlar 49. Crowfoot Satin weave	Cyanamid CYCOM 919
E	$((0,90)^k/0_3)_s$	8	1.57	Thornell 300, U/D Tape Kevlar 49, Crowfoot Satin Weave	Hexcell F263 Hexcell F151/1881
F	$((0,90)^k/0,90_2)_3$	6	1.47	Thornell 300. Plain Weave Kevlar 49, Crowfoot Satin Weave	Hexcell F263 Hexcell F161/188
G	$((9,90)^k/(0,90)'_s$	4	1.37	Thornell 300. 8-H-S Weaver Kevlar 49 Crowfoot Satin Weave	Hexcell F263 Hexcell F161/188

[*] Reprinted with permission from the Royal Aeronautical Society.

tests are generally required. In particular, to determine the threshold kinetic energy at which delamination is initiated requires accurate control of the impact velocity as well as the pre-load, thus requiring a large number of tests. As an alternative approach to the measurement scheme mentioned above, it has been observed by Sjoblom [7] that when the kinetic energy of the impactor is just above the threshold value, a drop in the measured contact force is noted. This force threshold P_i can be measured during a single impact experiment and has been shown to follow an equation of the type,

$$P_i = ch^{1.5}$$

where h is the laminate thickness and c an empirically determined constant.

As previously mentioned, transverse matrix cracking occurs in various types of composites, particularly PMCs (polymer matrix composites) when subjected to impact loads. Such cracks have been observed in the case of static loads, for example, by Broutman [5]. For the case of impact loads, the crack spacing for the front and back surfaces of impacted cross-ply glass-epoxy composite plates has been studied by Takeda et al. [8]. Specifically, the mean transverse crack distance, denoted by MTCD, has been introduced as a metric as measured on the front and back face surfaces of impacted composite plates for different types of projectiles. It has been observed that the threshold level for the development of transverse cracks appears to be independent of the impactor nose shape. Once, however, the threshold velocity is reached, the transverse crack spacing decreases rapidly with increase in projectile velocity. It appears that at some projectile velocity, the MTCD approaches a constant value, a saturation limit, this value being of the order of 3 mm for the case of $(0_5/90_5/0_5)$ cross-ply glass-epoxy composites. The above trends have been observed for impactors of arbitrary nose shape and length. Typical results obtained for the MTCD for blunt-nosed projectiles of two different lengths impacting $(0_5/90_5/0_5)$ targets are in Figure 5.7.

In addition to the surface crack studies discussed above, studies of internal cracks have been made using such tools as the scanning electron microscope (SEM). Post-impact interrogation of impacted targets suggests that where there is no delamination evident, transverse cracks grow perpendicular to the lamina interfaces. However, at lamina interfaces where a delamination crack exists, the transverse cracks appear to grow obliquely to the lamina interface [9]. Also, a

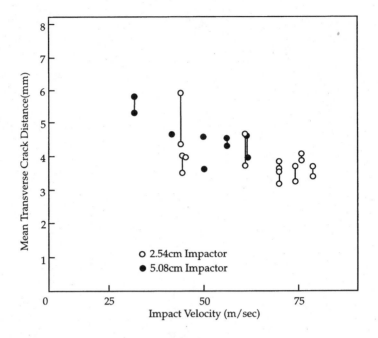

Figure 5.7   Mean transverse crack distance as a function of impact velocity for crossply glass-epoxy composites [8]. (Reprinted with permission from Chapman & Hall.)

higher density of transverse cracks has been found in regions of high fiber density as opposed to resin-rich regions. It also appears, based upon experimental evidence, that transverse cracks and delamination may occur simultaneously and be interactive.

A summary of the factors important to the initiation, as well as the control of damage for impacted composites, has been studied by Dorey [10]. For the case of carbon fiber reinforced plastics subjected to transverse impact, the damage occurring was found to depend upon the incident energy and momentum, the material properties, and geometry. Calculations of the energies involved in the damage process include [11]

- delamination energy $= \frac{2}{9}(\tau^2/E)(wl^3/t)$
- flexural fracture energy $= \frac{1}{18}(\sigma_f^2/E)(wlt)$
- penetration energy $= \pi\gamma td$

where

τ = interlaminar shear strength

σ_f = flexural strength of the composite

E = Young's modulus of the composite

γ = through the thickness fracture energy

d = diameter of the projectile

w, l, t = width, length and thickness of specimen, respectively

The type of failure occurring was found to depend upon the magnitudes of t and s and the span-to-depth ratio. In addition, penetration of the target was found to be dependent upon the incident impact energy as well as the size and shape of the impactor with perforation more likely to occur for small masses traveling at high velocities.

The above discussion is based upon linear elastic modeling. Another important question arises as to what happens when a critical level of damage is attained. For example, in brittle systems, crack growth occurs until the stored energy is dissipated, while in deformable materials, the energy may be dissipated without damage growth. Since most resin matrix systems are reinforced with fibers that exhibit a linear elastic response, improvements in impact resistance and damage tolerance are dependent upon improved matrix toughness. In addition, the work to failure may play an important role as a measure of the compressive strength after impact which can be expressed mathematically by

$$CSAI = K \int_0^{e_f} \sigma(\varepsilon) \, d\varepsilon$$

where

CSAI = compressive strength after impact

K = correlation constant

e_f = strain at failure

σ = stress

The capacity of a composite material system to tolerate a particular damage state (type, degree and size of damage) without seriously affecting its required functional characteristics appears to be the key issue within the damage tolerance design concepts and methodologies. Thus a need exists to identify some quantitative measures in order to determine how serious an effect is, serious in terms of short duration and/or of the residual service life. Questions of this type enter into the domain of reliability and life-prediction methodologies. Another consideration that warrants attention is the fact that a damage state occurring within a composite system may be more serious in one application than in another. Therefore, it appears that the damage tolerance concept is embedded within the following two broader areas.

1. correlations between damage states and corresponding residual material and structural characteristics;
2. correlations between residual characteristics and application-specific limiting parameters. These limiting parameters may depend upon a number of considerations such as reliability, safety, long-term service life, potential for further damage growth and other service factors such as environmental conditioning.

The limiting parameters mentioned here as yet to be defined and quantified for various practical situations, and therefore, it may not be possible to provide the necessary answers that designers may be hoping for to incorporate damage tolerance into design. The following sections focus upon ideas related to the residual characteristics of damaged composite systems.

5.3.2 Residual Strength/Stiffness

For damage-tolerant structures, data on the retained residual strength capacity of such systems when subjected to impact damage is essential. Figure 5.8 describes the various elements which may be important to developing such a structural design methodology.

The ability of a structure to sustain impact loads necessitates that the structure be capable of operating at load levels which will not

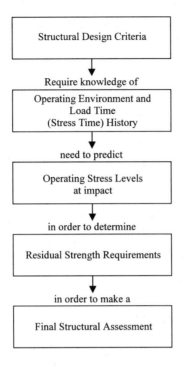

Figure 5.8   Flow diagram for structural assessment process.

1. exceed the required residual strength (stiffness) capability of the structure;
2. exceed specified cyclic loading levels.

The first condition can be related to either the strength or the stiffness-based critical design methodology. A review of the literature indicates that the residual strength capacity of a material/structural element has been studied in greater detail than has the residual stiffness; thus the focus is on residual strength data. This strength assessment can be further broken down into studies on the tensile, compressive, flexure, and shear modes of loading.

It is important to note that in connection with strength/stiffness degradation studies of material/structural elements subjected to impact loads, these elements can be in an unloaded or a preloaded state at the time of the impact event. This factor can drastically change the retained residual strength of the element. As an example, a panel element subjected to an applied tensile load (a pre-loading state) at impact is shown in Figure 5.9 with the scenario of possible events that would follow.

The resultant residual strength of the damaged panel can be depicted schematically as shown Figure 5.10. For small impact

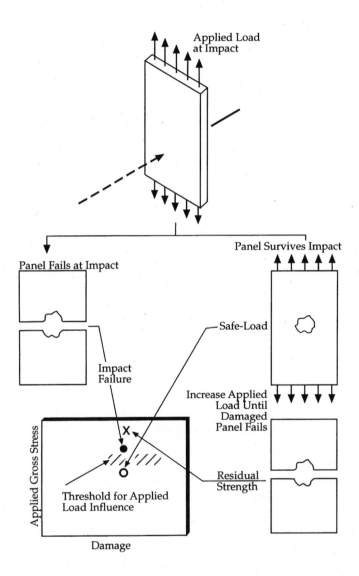

Figure 5.9 Strength degradation of tensile panels due to ballistic impact [17]. (Copyright ASTM. Reprinted with permission.)

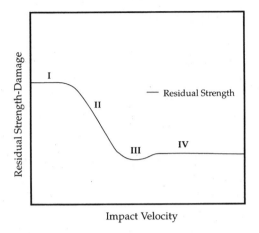

Figure 5.10 Post impact residual strength vs. impact (schematic).

velocities, as noted by Region I, and for which no damage is detected, no strength degradation is observed. Once damage is detected, as depicted by Region II, and physically represented by matrix cracking/delamination, the tensile strength is rapidly reduced. The maximum reduction in strength generally coincides with the maximum damage size and is depicted by Region III. At higher impact velocities, a complete perforation of the plate element occurs with a hole produced for which the residual strength then remains constant for increasing impact velocities. It should be noted, however, that the results presented here are dependent upon the velocity and the mass of the impacting object, that is, whether the object is of high or low mass. A matrix of possible impact events involving both a high/low velocity and a high/low mass is shown in Table 5.2. Most impact events which occur in practice fall into either the low-velocity/high-mass or high-velocity/low-mass categories. For example, the impact of a soft object (0.177 in RTV cylinder) impacting a boron/aluminum (B/Al) plate at velocities of up to 1500 ft/s may produce the same effect as a hard object (0.177 in steel ball) impacting a B/Al plate at a velocity of 100 ft/s [12]. Plate thickness and material can also affect the above results.

The assessment of residual strength with impact velocity/energy, as mentioned, has been studied for tension, compression, flexure, and shear modes of testing. Among the control parameters investigated have been

TABLE 5.2   Possible Impact Events

Velocity/Mass	High	Low
High	X	*
Low	*	X

- matrix type (thermoset, thermoplastic);
- fiber type (continuous, chopped);
- composite type (prepreg, filament wound, woven mat).

In addition, other factors such as laminated material properties, laminate thickness, impactor size and shape, impact energy, support conditions, and preloading play an important role in post-impact strength assessment.

A discussion of residual strength studies follows along with a tabulation of mathematical models developed for the systems studied (Table 5.3). Some suggested correlations between the residual strength/stiffness and damage-tolerance concepts are listed below.

1. correlating impactor characteristics (shape, size, mass, stiffness etc.) with the damage state (type, degree and size of damage) for various composite systems (fiber types, matrix types, stacking sequence, shape and size);
2. correlating damage states (type, degree and size) with the corresponding residual strength/stiffness (tensile, compressive, flexural and shear) of various composite systems;
3. correlating impactor characteristics (shape, size, mass, stiffness etc.) directly with the residual material and/or structural characteristics such as residual strengths/stiffnesses and static as well as dynamic structural response.

Attempts made to predict correlations under the categories mentioned above include empirical, semi-empirical, analytical and numerical models. A few models are discussed as illustrative examples. It should be noted, however, that such models and correlations have yet to be translated into proper damage-tolerance design critera which can be useful to designers.

Investigator	Strength Reduction Equation	Parameters
Husman et al., 1975 [13]	$$\frac{\sigma_R}{\sigma_o} = \left[1 - \kappa \frac{\overline{W}_{KE}}{W_s}\right]^{\frac{1}{2}}$$	σ_o is a proportionality constant. κ is a proportionality constant. W_s is the area under stress-strain curve for an unnotched specimen. $\overline{W}_{KE}$ is the kinetic energy imparted to the plate specimen.
Caprino, 1983 [14]	$$\frac{\sigma_R}{\sigma_o} = (C_o/C)^m$$	C_o is the length of an existing defect in the material. C is a characteristic length equivalent to the size of a notch. m is a constant independent of material, type of laminate, shape of discontinuity.
Awerbuch and Hahn, 1976 [12]	$$\frac{\sigma_R}{\sigma_o} = \frac{1}{Y(C)}\left[\frac{C_o}{C_o + C}\right]^{1/2}$$	C_o is the length of an existing defect. C is a characteristic damage size. $Y(C)$ is a finite width correction factor.
Avva et al., 1986 [15]	$$\frac{\sigma_R}{\sigma_o} = \frac{1}{\left[\frac{2\kappa}{\alpha_o}(W - W_o) + 1\right]^{1/2}}$$	κ is a constant. α_o is a characteristic length. W is the kinetic energy for unit thickness. W_o is a threshold value of W.
Lal, 1983 [16]	$$\frac{\sigma_R}{\sigma_o} = \left\{\frac{C_o}{[C_o + I_F/2hG_{IC}]}\right\}^{1/2}$$	c_o is the flaw size in the material. I_F is the fiber breaking energy. h is the thickness of the plate. G_{IC} is the energy required to produce a unit fracture surface area.
Classical stress concentration result	$$\frac{\sigma_N^\infty}{\sigma_o} = \left\{\frac{C_o}{C + C_o}\right\}^{1/2}$$	C is a characteristic crack length. C_o is a characteristic length.

Tensile Strength   In developing models for determining the residual strength of impacted fiber composites, certain velocity regimes have been studied. The appropriate and unique definitions of these regimes have yet to emerge and a low-velocity impact regime for a particular impactor–target situation may in reality be a high-velocity regime for another target. The regimes are

- low velocity;
- intermediate velocity;
- high velocity.

The analytical approaches taken to modeling the above regimes are

- empirical;
- semi-empirical;
- analytical;
- numerical.

Within the domain of the velocity regimes and analytical approaches studied, the low velocity and empirical/semi-empirical models have received the most attention by investigators. One such semi-empirical model for evaluating the tensile residual strength of impacted composites is that due to Husman et al. [13]. The focus of this work is related to studies of

- hard particle damage;
- impact velocities less than perforation;
- impact velocities greater than perforation;
- influence of fiber reinforcement.

The approach taken is to develop a procedure for converting impact damage into an equivalent crack size in order to predict the residual strength using fracture mechanics principles.

The analysis is based upon consideration of an orthotropic plate with a slit $2c$ subjected to a uniform tensile stress $\bar{\sigma}$ applied at infinity. The critical strain energy release rate G_{Ic} for the plate is

$$G_{Ic} = K_{Ic}^2 \left\{ \left(\frac{\bar{S}_{11}\bar{S}_{22}}{2} \right) \left[\left(\frac{\bar{S}_{22}}{\bar{S}_{11}} \right)^{1/2} + \frac{2\bar{S}_{12} + \bar{S}_{66}}{2\bar{S}_{11}} \right] \right\}^{1/2} \tag{5.1}$$

here, K_{Ic} is the critical stress intensity factor and the $\bar{S}_{ij}$ are the orthotropic plate compliances. The value of K_{Ic} is given by

$$K_{Ic} = \bar{\sigma}\sqrt{\pi c} \tag{5.2}$$

substituting K_{Ic} into G_{Ic} one gets the expression

$$G_{Ic} = Ac\left(\bar{\sigma}^2\frac{\bar{S}_{22}}{2}\right) \tag{5.3}$$

The bracketed term in the above expression represents the energy necessary to break the specimen given by W_B, therefore

$$G_{Ic} = AcW_B \tag{5.4}$$

Assume a damage zone adjacent to a stress concentration that constitutes a volume of material necessary to stress the material to a critical level before fracture. A critical strain energy release rate can then be written in terms of an effective flaw size c_0 as

$$G_{Ic} = Ac_0W_s \tag{5.5}$$

where W_s represents the work done under the stress-strain curve for a statically loaded composite. Using c_0 as the damage zone, an effective crack length $(c + c_0)$ is assumed so that one can write

$$G_{Ic} = A(c + c_0)W_B \tag{5.6}$$

The residual strength σ_R can then be written in terms of the unflawed strength σ_0 [13]

$$\sigma_R = \sigma_0\sqrt{\frac{c_0}{c + c_0}} \tag{5.7}$$

It is advantageous to relate small hard-particle impact damage to damage introduced by an implanted crack of known dimensions inserted in a static tensile coupon. For velocities less than the perforation velocity, damage is assumed to be related to the kinetic energy imparted to the plate. Thus,

$$W_s - W_B = k\frac{W_{KE}}{V} \tag{5.8}$$

where W_{KE} is the kinetic energy imparted to the specimen and V is the volume over which the energy is dissipated. This volume can be assumed to be related to some characteristic surface area A_e which does not depend on the kinetic energy of impact, and the plate thickness, t. Therefore,

$$V = (A_e)t \tag{5.9}$$

rewriting the expression for $W_s - W_B$, as

$$W_B = W_s - K\bar{W}_{KE} \tag{5.10}$$

where $K = k/A_e$ and $\bar{W}_{KE} = W_{KE}/t$.

By combining equations (5.5), (5.6), (5.7), and (5.10) we can write the residual strength of the impact-damaged specimen as

$$\sigma_R = \sigma_0\sqrt{\frac{W_s - K\bar{W}_{KE}}{W_s}}$$

The above expression indicates that σ_R can be found by conducting two types of experiments:

- a static tension test on an unflawed specimen;
- a static tension test on a damaged specimen.

It should be noted that for sufficiently wide specimens K is independent of geometry; however, σ_R may be influenced by the specimen boundary conditions and the material stacking sequence of the specimen. One of the key parameters in the expression for σ_R is W_S. An analytical method for finding W_S is to define the laminate properties, stacking sequence and ply orientation as described in Figure 5.11. Some results using the technique described above demonstrating the effect of fiber type on the impact residual strength are shown in Figure 5.12.

Energy Balance Model   An energy dissipation model based upon fracture mechanics principles has been advanced to predict the

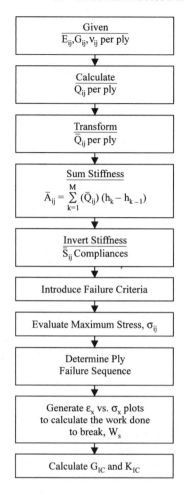

Figure 5.11   Flow diagram for calculating fracture toughness.

residual strength of transversely impacted composite laminates. In the model development it is noted that the impact damage in composites consists primarily of

· matrix cracking;
· fiber/matrix debonding;
· fiber fracture;
· peeling and/or pulling out of broken fibers.

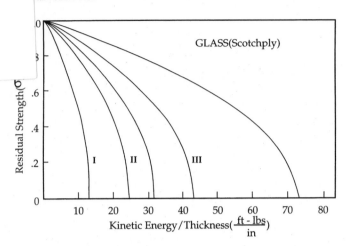

Figure 5.12 Residual strength as a function of kinetic impact energy per unit thickness [17]. (Copyright ASTM. Reprinted with permission.)

For low-velocity impacts, the first two damage mechanisms represent the principal internal damage modes in the composite laminate. The concept of an energy balance model is introduced in which the following quantities are calculated,

- the energy absorbed by delaminations, I_{Del};
- the net energy absorbed by the target, I_t;
- the energy used for reducing the strength of the structure due to fiber breakage, I_f.

In order to evaluate fiber breakage energy, the load-deflection curve given in Figure 5.13 can be used. Specifically, the area bounded by the curves represents the net energy absorbed by the target, that is I_t. The fiber breakage energy can be written in terms of the area bounded by the curves shown in Figure 5.14, and is given by

$$I_f = I_t - I_{Del} \qquad (5.11)$$

A plot of the respective energy losses is shown in Figure 5.14. In order to calculate the residual strength, the energy required to produce a through the thickness unit area of fracture surface energy is given by

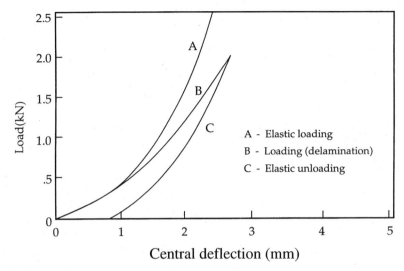

Figure 5.13 Elastic loading, progressive delamination and elastic unloading [16]. (Reprinted with permission.)

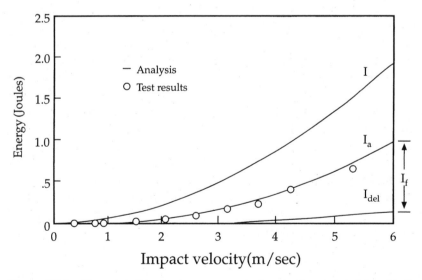

Figure 5.14 Computed impact energy, net absorbed energy, delamination energy, and fiber-breakage energy with impact velocity [16]. (Reprinted with permission.)

$$\sigma G_{\text{IC}} = K_{\text{IC}}^2 \left\{ \left(\frac{\bar{S}_{11}\bar{S}_{22}}{2} \right) \left[\left(\frac{\bar{S}_{22}}{\bar{S}_{11}} \right)^{1/2} + \frac{2\bar{S}_{12} + \bar{S}_{66}}{2\bar{S}_{11}} \right] \right\}^{1/2} \qquad (5.12)$$

Following the arguments presented previously in deriving Eq. (5.7) and noting that the equivalent slit length c can be obtained from the strain energy release rate and the fiber breakage energy, then

$$c = \frac{I_f}{2hG_{\text{IC}}}$$

and the residual strength after impact can be expressed as

$$\frac{\sigma_R}{\sigma_0} = \left\{ \frac{c_0}{c_0 + \dfrac{I_f}{2hG_{\text{IC}}}} \right\}^{1/2} \qquad (5.13)$$

Results obtained by Lal [16] for T300/5208 graphite-epoxy composites are shown in Figure 5.15. These results indicate that the residual strength of the impact damaged specimens is reduced with increase in the impact velocity; the strength reduction approaching that of a laminate with a drilled hole equal in size to the impactor. The analysis suggests that impact resistance is enhanced by development of composite materials that have a higher strain to failure.

A summary of predictive equations for strength reduction after transverse impact loading of composite plates is included in Table 5.3 along with appropriate references. It is interesting to observe that predictions derived from classical stress concentration analysis have a form similar to the residual strength formulae derived for the models discussed.

Compression Strength   Impact-induced delamination can lead to a number of modes of failure including gross compression, local or global buckling, and delamination growth. The last event, for a given geometry and material system, has been assumed to occur when the strain-energy release rate exceeds the required energy level to induce additional delamination.

In general, compression testing of composites has involved tests conducted on flat panel specimens in both an unloaded and pre-loaded state. Further work has been performed on stringer reinforced

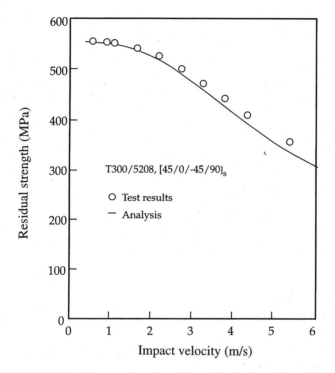

Figure 5.15 Residual strength comparisons for impact damaged specimens [16]. (Reprinted with permission.)

panels with different numbers and types of stringers in both the unloaded and loaded state. Early work reporting on the failure mechanisms and damage propagation in compression-loaded panels subjected to impact loads are described by Rhodes et al. [18]. In the experimental investigation cited, aluminum spheres were projected at graphite epoxy targets using a compressed air gun at four different velocities ranging from 35 m/s to 125 m/s. Damage was classified into

- surface damage;
- interior damage.

Surface damage was identified through visual inspection while interior damage was detected by sectioning impacted laminates and ultrasonic scanning. The effect of the resultant impact damage on the compression strength of orthotropic and quasi-isotropic laminates is shown in Figure 5.16. A plot of the applied axial strain in the specimen due to the applied compressive load at the time of impact versus the impactor velocity is shown. The curves as shown represent the threshold level between fail safe and failed specimens. As noted from the tests conducted, there appears to be little difference in the results obtained for the tape and woven graphite materials tested. Some specimens however were observed to fail catastrophically, depending upon the applied compressive preload. It was observed that impacted specimens with no applied load had local damage regions smaller than that of specimens with an applied compressive load. Further, it was noted that as the applied axial strain increased, there was a trend toward coupling between the applied axial strain and the out-of-plane deformation caused by impact loading, resulting in increased

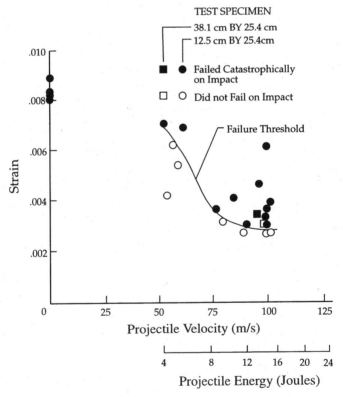

Figure 5.16 Failure threshold of impact-damaged composite [18]. (Reprinted with permission from Society of Plastics Engineers Inc.)

damage size. It was also observed that for specimens which failed catastrophically under impact such specimens were observed to have local out-of-plane deformations with a wave length of the order of the plate thickness. Once impact damage was initiated, the damage could propagate in one of three ways, these being

- delamination;
- coupling between axial loading and lateral deformations;
- local shear propagation.

Each of these modes can cause a significant reduction in compressive strength.

A follow-on study investigating the effects of impact damage and geometric imperfections on the retained structural length of graphite-epoxy panels has been reported on by Williams et al. [19]. In this study, a computer algorithm was used in tailoring the geometric dimensions and ply configuration to the loading and constraint conditions. One and three bay frame configurations were tested, with both front and back surface damage observed, and interior damage inspected ultrasonically. Results for specimens impacted at a prescribed axial strain are shown in Figure 5.17. These data are plotted as strain at impact versus impactor kinetic energy. The trend in the data indicates that the compressive strength of the specimens tested is affected at higher impact energies. Several damage-propagation mechanisms have been observed including (1) delamination, (2) axial load–lateral deformation coupling, and (3) local shear. It is also observed that the closer the delaminations are to the surface, a lower load is required to initiate local buckling. This type of failure propagation also occurs for panels subjected to cyclic loading. The second failure mode is observed to occur in impacted panels under load at strains near the failure threshold value. The third failure mode, local shear failure, involves transverse shear failures with short wave length.

Residual strength tests were also conducted on specimens with open circular holes and compared with impacted specimens. Test data indicate that specimens with a 0.16 cm diameter hole failed away from the hole while specimens with a larger hole failed at threshold failure strains less than that of controlled damaged specimens.

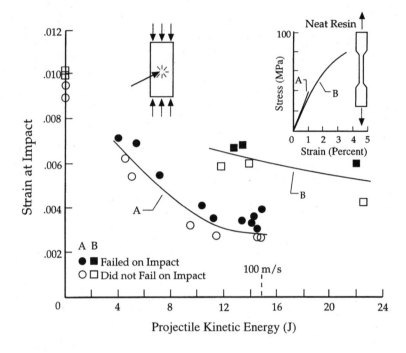

Figure 5.17 Effect of matrix on damage tolerance [19].

The effect of stiffeners in panel structures, including blade and hat stiffeners, has also been studied. Results for these panels indicate that the effects of both low velocity impact damage and circular holes can severely degrade the compressive strength of heavily-loaded panels.

Another key element in strength degradation/reduction is that due to the influence of the composite matrix. A comparison of two different matrix types, having the stress-strain curves shown in the inset of Figure 5.17, shows that the failure threshold strain is higher for material B. Material B is observed to have a similar tensile modulus to Material A while having twice the ultimate strength and with a strain to failure four times that of Material A. It is noted that other properties can also affect composite damage tolerance as well, including

- shear modulus and strength;
- resin content;
- strain-rate sensitivity.

The effect of specimen size on the buckling strains of composite laminates made of graphite-epoxy (T300/5208) subjected to low-velocity impact have been studied by Avva [20]. Two composite thicknesses, a 16-ply and a 32-ply, with stacking sequence ($\pm 45/0/90$) were subjected to low-velocity impact using an aluminum sphere impactor. The total number of specimens tested in each series of tests was between 15 and 20. Results have been plotted in terms of a pre-stress (pre-strain) or residual stress (strain) as ordinate versus impact energy.

An evaluation of the compressive strength after impact using the amount of damage as measured by a non destructive evaluation (NDE) was investigated by Hirschbuehler [21]. Both the influence of neat resin modulus and strain or work to failure on compressive strength were studied. The latter effort has been graphically depicted through plots of compressive strength versus G_{Ic}. Results obtained indicate that the compressive strength increases as G_{Ic} increases; also, the compressive strength decreases as the impact damage area increases. It appears from these studies that the resin strain to failure is a controlling factor in composite impact performance.

Studies by Leach et al. [22] on the post-impact compressive strength of carbon fiber/PEEK composites indicate two types of failure modes:

1. For impact energies less than 15 J macro/global buckling occurs.
2. For impacted energies greater than 15 J a combination of local buckling and shear occurs.

In addition, the following observations on compressive loaded specimens were noted:

1. There is no evidence for delamination growth for deformations up to 0.60%.
2. Compressive failure deformation occurred for complete penetration of the specimen.

A qualitative representation of the effect of delamination size on the failure modes can be observed in Figure 5.18. From observations made from this figure, Table 5.4 can be constructed.

In summary, it can be observed that since the internal damage pattern associated with a given impact event results predominantly in

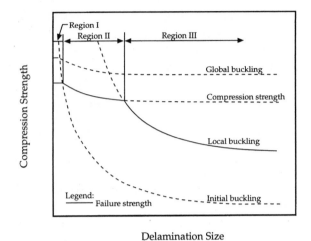

Delamination Size

Figure 5.18   Schematic illustration of the effect of delaminationn size on the compressive failure mode.

delaminations between plies, studies of the delamination size and growth are important parameters for characterizing the post-impact strength of composites.

Residual Bending Strength   Post-impact studies on the retained residual strength and stiffness of composite specimens (graphite, Kevlar, and glass-epoxy) in flexure have been reported by Malvern et al. [23]. Results obtained for the residual strength of impact-damaged plate specimens as measured with respect to the undamaged plates have been tabulated and graphically depicted in Figures 5.19 and 5.20. It was observed that the strength degraded rapidly for small delamination areas while the stiffness stayed essentially constant, until a threshold delamination area was reached, after which there appeared

TABLE 5.4   Effect of Delamination Size

Region	Delamination Size	Buckling Strength	Compression Strength	Failure Mode
I	Small	High	Low	Compression
II	Intermediate	Low	High	Buckling
III	Large	Low	High	Local buckling

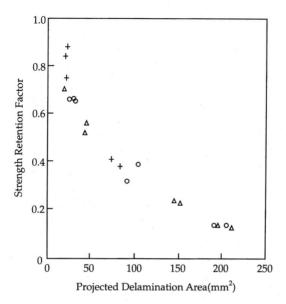

Figure 5.19 Strength retention factor vs. delamination area for Kevlar-epoxy [23].

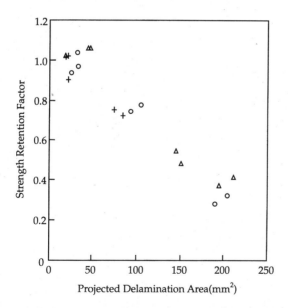

Figure 5.20 Stiffness retention factor vs. delamination area for Kevlar-epoxy [23].

to be a linear loss in stiffness. Other studies on the bending strength of composites have been reported by Cantwell and Morton [24] and Rotem [25]. In the former study [26] a model based on a simple failure criterion and classical laminate theory was used to predict the bending strength after impact. Reasonable agreement was found in comparing results between theory and experiment. Flexural impact damage tolerance studies by Rotem [25] were conducted on brittle and ductile laminates made of graphite-epoxy and glass-epoxy respectively. A correlation study of the damage done to the beam during impact by measuring the residual modulus and strength was conducted. An assessment of the factors causing this result has not been made, that is, how damage occurs, whether from bending or contact forces but rather on the resultant effects. Strength degradation as a function of impact energy has been shown in Figures 5.21 and 5.22. It is observed that both ductile and brittle specimens having some visible damage retain an average of 90% of the residual strength of unimpacted specimens. Changes in the loading modulus of impacted ductile and brittle specimens are however different. For a ductile material, the stiffness decreases with increase in impact energy while for the brittle material, there is no apparent change in impact energy up to a point where suddenly there is a total loss of stiffness.

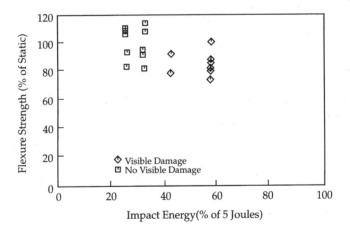

Figure 5.21 Strength degradation of glass-epoxy laminates due to impact energy [25]. (Reprinted with permission from the Society for the Advancement of Material and Process Engineering.)

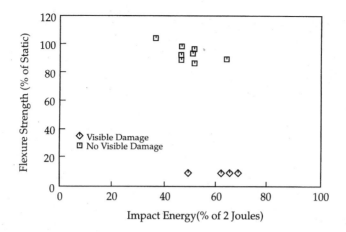

Figure 5.22   Strength degradation of glass-epoxy laminates with impact [25]. (Reprinted with permission from the Society for the Advancement of Material and Process Engineering.)

Residual Shear Strength   Residual shear strength studies for impacted carbon, Kevlar, glass, and hybrid laminates have been studied by Dorey [10] and Llorente [27]. Dorey used short-beam shear tests to evaluate critical shear effects. Linear elastic fracture mechanics was used as the basis for calculating the critical shear stress t, expressed analytically by

$$t \sim K_c(\pi a)^{-1/2} \tag{5.14}$$

where a is the crack length and K_c the critical stress intensity factor. An important concept introduced was that the crack length is proportional to the square root of the delaminated area A; thus

$$t \sim K_c(A)^{-1/4} \tag{5.15}$$

Llorente [27] observed that interlaminar shear stresses cause inter-ply delamination and that such results correlate well with fracture stress analysis. Results of tests conducted on short-beam shear tests were used to establish data on the interlaminar shear strength of the materials tested.

5.3.3 Residual Dynamic Response

Vibrations As discussed earlier, delamination, matrix cracking and fiber fracture represent the primary damage that is induced in impacted specimens. One measure for interpretation of the damage state of an impact-damaged component can be considered to be associated with the residual dynamic response of the material/structure. Recent experimental evidence suggests that delamination represents a dominant mechanism in flexural stiffness loss in transversely impacted composites. Since frequency response and stiffness are interrelated, the effect of delamination as measured by the natural frequencies and damping properties of composites appears to be a potentially useful tool for evaluating impact damage in material/structural specimens. Specifically, an investigation of T 300/934 graphite-epoxy test specimens consisting of 20 plies arranged in a cross-ply $(0/90)_{5s}$ configuration has been reported by Grady and Meyn [28]. Delamination damage was introduced by impacting cantilever beam specimens with rubber impactors. The extent of delamination has been measured by using (1) visual observation along the specimen edge, and (2) ultrasonic scanning. Data were obtained and analyzed for both simulated impact damage and impact induced damage. Results as shown in Figure 5.23 suggest that delamination appears to be the dominant mechanism for flexural stiffness loss leading to a corresponding drop in the natural frequency of the material/structure [28].

Wave Propagation A method for prediction of damage development in composite materials has been suggested as stress wave/material interaction induced by dynamic load redistribution. The basis for the characterization and prediction of impact damage development is the use of acousto-ultrasonic (AU) measurements [29]. Specifically, damage development corresponding to AU-measured response is considered to be related to the fiber–matrix architecture and the material properties. Cross ply laminates have been used to study the resultant damage occurring from impact in materials with and without prior induced damage.

In addition to the characterization/prediction of impact-induced damage, the AU technique can be used to locate impact damage in composites. Such studies have been performed on graphite-epoxy quasi-isotropic laminates having a ply sequence of $(45/0/-45/90)_{6s}$. Damage was introduced into the specimens by impacting the speci-

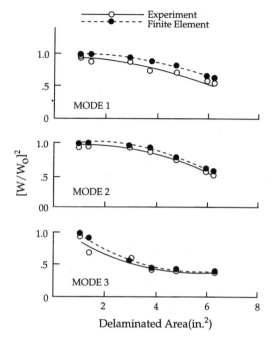

Figure 5.23 Variation of natural frequency with delamination area [28]. (Reprinted with permission from ASME.)

men with an aluminum impactor [30]. Parameters useful to locate composite damage include changes observed in peak amplitude, RMS amplitude, and energy distribution in the material. As an example of the energy distribution parameter, Figure 5.24 shows the energy transmitted in unimpacted and impacted regions. The principle for detection in using this parameter is the energy transmission of the wave form itself. Thus small differences in peak amplitude result in greater differences in energy transfer. By operating at transmission frequencies at or near maximum transducer sensitivity, higher amplitude waves can be obtained. Since damage scatters waves and diminishes the energy content in the wave form, such tests are extremely useful in damage detection.

5.4 EFFECTS OF MATERIAL PROPERTIES/PARAMETERS

As discussed earlier, the damage in composites represents a combination of events including delamination, matrix cracking and fiber

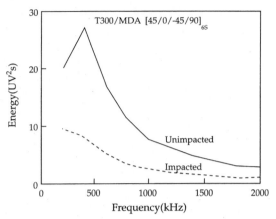

Figure 5.24 Correlations of imparted wave energy and impact frequency for T300/MDA laminates [30]. (Reprinted with permission from Technomic Publishing Co., Inc.)

breakage. The first two types of damage are related to the matrix properties while the last is related to fiber properties.

Matrix In general, studies on polymeric composites indicate that composite performance when subjected to impact damage can be enhanced through improvements in the toughness of the resins systems [3]. This mechanical property represents a measure of the ability of the material to absorb strain energy, resist shear cracking, and reduce the effects of stress concentrations [31,32]. It has been observed that thermoplastic matrix composites exhibit relatively higher toughness [3]. In general thermoplastic composites produce less matrix cracking and exhibit less extensive damage. Also, it is observed that resins which have a higher strain to failure are able to resist higher impact loads [21,33]. In addition to sustaining higher impact loads, toughened systems also demonstrate a higher residual compression strength after impact, since there is less delamination.

Fiber Another important aspect of damage control is the role of the fiber in improving the damage tolerance of the composite. As an example, for equivalent impact energies, the ability of the composite to absorb energy results in less fiber breakage and higher residual strengths. Some results reported on the performance of such composites have been reported by Cantwell et al. [34,35].

K.J.abbott@qmul.ac.uk

l. wright@millmily.com

millam-patel@mcfair.com

Fiber-Matrix Interface The fiber-matrix interface is as important as the matrix and fiber constituents in the control of composite damage. For short-fiber composites a critical fiber length embedded in a matrix is necessary to transfer load from the matrix to the fiber. For continuous fiber composites, the introduction of a material interface can increase the impact resistance of the composites. For example, studies by Peiffer [36] indicate that for the case of glass-epoxy composites, the introduction of a rubber-like interface can be useful in improving the damage tolerance of composites.

Stacking Sequence The ply stacking sequence in laminated composites has been shown to play an important role in the damage tolerance of composite [37]. In particular, depending upon whether a rigid impactor strikes a rigid or flexible target, the damage initiated may represent a front to back or back to front damage development. Experimental evidence indicates that target stiffness represents a dominant parameter and controls the mode of fracture. For example, flexible targets on impact generate large tensile stresses in the lowest ply, resulting in failure initiation at the fiber-matrix interface. Cracks occurring at the interface are propagated and transmitted from back to front resulting in ply delaminations.

At impact energy levels which result in local damage initiation, damage occurs from front to back, with damage spread from high tensile stresses generated in the vicinity of the impactor. Such stresses are sufficiently large to cause failure at the fiber-matrix interfaces resulting in crack transmission progressing from front to back, with corresponding delamination. A schematic of this damage progression is shown in Figure 5.25.

Studies by Cantwell [38] on impacted CFRP Composites targets of varying thickness, with the same stacking sequence, indicate that the projected delamination area varies linearly with the impacted kinetic energy of the projectile. This result follows that observed by other investigators including Sierakowski et al. [1], Takeda et al. [8], and Malvern et al. [39,23]. The projected delamination area as a function of the kinetic energy has been found to vary nonlinearly with increase in target thickness. It has been discussed in the literature that if a relationship between the total delamination area and impacted energy exists, then the results from tests of different numbers of stacking sequences can be plotted on one curve. Experimental results reported by Malvern et al. [39] and shown in Figure 5.26 support this assumption.

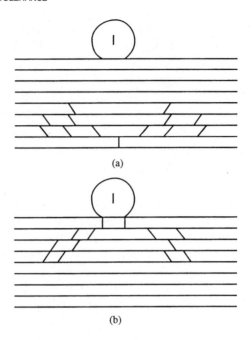

(a)

(b)

Figure 5.25 Schematic representation of damage progression due to (a) flexural stress and (b) contact stresses.

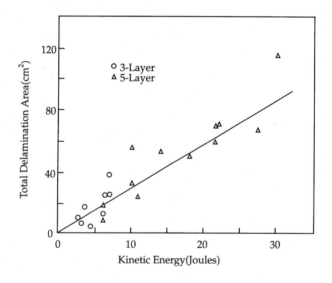

Figure 5.26 Total delamination area vs. imparted kinetic for graphite-epoxy laminates [40].

The role of stacking sequence for the same thickness composite target has been studied by Ross and Sierakowski [37] and Cristescu et al. [41]. A systematic study of cross-ply glass-epoxy composite plates indicates that for the system studied, a sequential delamination process occurs which plays an important role in dispersing the absorbed energy. This process is described in detail by Malvern et al. [39] and is based upon a sequential delamination mechanism initiated by a generator strip. A brief review of this process follows. At low to moderate impact speeds occurring from impact of a blunt-ended projectile against a cross-ply composite plate, two parallel cracks are produced through the thickness in the first lamina at a distance equivalent to the impactor. This first strip, identified by the letters AA and BB in Figure 5.27, presses against the second lamina until further penetration or rebound occurs. This process is repeated through successive laminae or until the energy necessary to drive the propagating delamination crack is exhausted. The timing of the cracks at the interfaces between successive plies appears to occur at approximately the same time. Also, for the cross-ply system studied, the length of the generator strip and the extent of the delamination are found to be equivalent. This is illustrated by area A_1 in Figure 5.27. Further sequential delaminations follow this pattern. The theoretical model advanced correlates well with documented experimental data as discussed by Cristescu et al. [41].

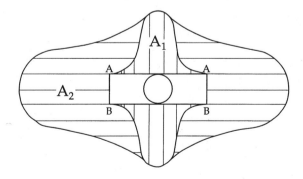

Figure 5.27 Schematic sequential delamination patterns with generator strip [39].

TABLE 5.5 Summary of Investigations Discussed in Chapter 5

Reference	Target Material	Target Stacking Sequence	Tension	Compression	Flexure	Shear
Adsit and Waszczak [42]	AS/350	$[10/\pm 45/0]$	Reduction in tensile strength related to size of damage and energy of threat, 50% reduction	Honeycomb core w/composite facings is decreased as energy of impact is increased		
Avery and Porter [43]	Boron/epoxy Graphite/epoxy	$[0_m/(\pm 45)_{m_s}]$	Failure of preloaded panels in tension occurs at levels significantly below the residual tensile strength			
Avva [20]	1300/5208	$[\pm 45/0/90]_{m_s}$		Strain ratio vs. impact energy appears to give meaningful information		
Bishop [31]	APCI (carbon/ PEEK)	$[\pm 45/0_m/\pm 45_m]$	• Carbon/PEEK more notch sensitive than carbon/epoxy in tension. • Front surface damage of carbon/PEEK results in higher residual and tensile strength than carbon/epoxy	Residual compressive strength of carbon/PEEK consistently greater than carbon/epoxy after impact		
Butcher [32]	Courtaulds type 2(HTS) carbon fiber/epoxy	(0_{ms}) unidirectional	Angled impacts have more serious weakening effect in unstressed specimens	Toughness		
Butcher and Fernback [44]	Courtaulds type 2(HTS) carbon carbon/epoxy	unidirectional	Toughness			

Author	Material	Layup			
Cantwell/Morton [24]	CRRP	$[0/90/0/\pm45/0]_s$	Reduction in tensile strength linear up to 4J		Reduction in flexure strength linear up to J4J, 60% reduction
Cantwell et al. [35]	AS4/3501-6	$[0/(\pm45)_2]_s$	Increased strain to failure increases residual tensile performance by up to 100%	Increased strain in failure can increase individual residual strength by up to 30%	High-strain composites had better residual flexural strength
Cantwell/Morton [45]	Gratil XA-S/Cita-Geigy BSL914C	$[(\pm45)_m]_s$	Up to 50% reduction in residual tensile strength		
Cantwell et al. [34]	CFRP woven and non-woven	$[\pm45/0_2/\pm45/0_2]_s$	Up to 50% reduction in tensile strength of non-woven laminates Residual tensile strength of mixed woven laminates greater than non-woven	Up to 50% compressive strength reduction in non-woven laminates vs. mixed woven	
Clerico et al. [46]	Graphite/epoxy Graphite/bismaleimide	$[\pm45/(0/90)_2]_2$ $+45/90/-45/0/+45/90]_s$	Reduction in tensile strength by up to 50%	Reduction in compression strength by over 50%	
Curtis and Bishop [47]	CFRP woven and non-woven	$[0/90], [\pm45]$ $[0/90/\pm45], [0/\pm45]$	Improved residual Tensile strength for woven composites	Improved residual compressive strength for woven composites	
Demuts et al. [48]	T30/5208 A5/3501-5 A6/2220-3	$[\pm45/90(\pm45/0_2)_2]_s$		Improved residual compressive strength for Cycom 907 vs. baseline epoxy	Cycom 907 used as benchmark for toughnes
Evans and Masters [11]	Graphite/epoxy (Cycom 907)	$[(\pm45/0/90/090)_2 \pm45/0/90/\pm45]_s$		Improved residual compressive strength for Cycom 907 vs baseline epoxy	Cycom 907 used as benchmark for toughness

(Continued on next page)

TABLE 5.5 Summary of Investigations Discussed in Chapter 5 (*continued*)

Reference	Target Material	Target Stacking Sequence	Tension	Compression	Flexure	Shear
Gandhe and Griffin [49]	T300/934	$[0/90_4]_s$	Interleaved laminates show higher residual tensile strengths after impact.			
Griffin [50]	Fiber/resin AS4/3502 Celion A536 2220-1 HST 584 1806 5245C 974 HST-7	Quasi-isotropic		Residual compression strength of toughened vs. untoughened matrices show improved strength		
Lal [16]	T300/5208	$[45/0/\text{-}45/90]_s$	• Residual tensile strength reduced by 40% • Prediction of residual strength made			
Llorente and Mar [51]	Graphite/epoxy Gelion 6K/Epon 9400 9450	$[\pm 30/90]_s$ $[\pm 60/0]_s$		Compression strength reduced by impact damage for filament wound system		
Malvern et al. [39]	Glass/epoxy Kevlor/epoxy Graphite/epoxy	$[0/90]_{ms}$			Strength degradation greater than stiffness degradation	
O'Kane and Benham [4]	CRFP/epoxy Kevlar 49/epoxy CFRP/Kelvar 40/epoxy	$[\pm 45/0_3]$ $[(\pm 45/(0/90)_2 \pm 45)]$ $[(0/90)^k/(0/90)]_s$ etc	Reduced tensile strength	Reduced compression strength	Reduced flexural strength	Residual shear can be reduced from 40% to 80%, depending upon

Ramkumar et al. [52]	T300/V-378A Graphite/bismaleimide	$[0]_{16}$ $[90]_{16}$ $[\pm45]_{43}$ $[0/\pm45/90]_{25,45,5}$		Residual compressive strength reduced by over 60%
Sankar and Sun [6]	Graphite/Epoxy	$[(0_2/90_2/0_2)]_s$	Initial stress (1/3 ultimate) reduces tensile strength by about 33%	
Sharma [53]	T300/5208 T300/934	$[\pm45/0]_s$ $[\pm45/90/0]_s$ $[90/\pm45/0]_s$	Residual stiffness reduced by up to 65%	Residual strength reduced up to 50%
Takamatsu et al. [54]	T700/3620 T300/3601 AS4/PEEK			Compression residual strength most severe design condition after impact
Verpoest et al. [55]	Scotchply 1003	$[(0/90)_2/0]_s$	Tensile strength reduction of 50% at 10J of energy	Residual compressivie strength reduction faster than tension
Williams and Rhodes [56]	Graphite/epoxy	$[\pm45/0_2/\pm45/0/90_2]_{2s}$		Residual compressive strength reduced by prestrain

5.5   CONCLUDING REMARKS

On the basis of a number of reported studies, the following general trends can be noted. (1) The tensile and compressive residual strengths decrease with increase in the extent of damage, with the correlation dominantly nonlinear. (2) The delamination damage size increases with increase in the impact energy and increase in impactor velocity until a threshold value is reached at perforation. Two types of threshold values are observed, one with respect to impact energy, one with respect to impact velocity; these represent a lower bound value for visible damage and an upper bound value for the maximum damage zone size. (3) Actual correlations for the above measures of damage may depend upon a number of material and geometrical parameters such as fiber, matrix and fiber-matrix bond strength, flaw sensitivity, fracture toughness, stacking sequence, interlaminar strength, boundary conditions, inherent material damping and other factors.

A summary of the various investigations discussed in this chapter has been included in Table 5.5.

5.6   REFERENCES

1. Sierakowski, R.L., Malvern, L.E., and Ross, C.A. (1976) *Proceedings of the 105th AIME Meeting*, Las Vegas, Nevada.

2. Liu, Dashin (1987) "Delamination in stitched and nonstitched composite plates subjected to low-velocity impact," *Proceedings of the American Society for Composites*, University of Delaware, September, pp. 147–155.

3. Dorey, G., Bishop, S., and Curtis, P. (1985) "On the impact performance of carbon fibre laminates with epoxy and PEEK matrices," *Comp. Science Technol.*, **23**(3), 221–237.

4. O'Kane, B.A.A. and Benham, P.P. (1986) "Damage thresholds for low velocity impact on aircraft structural composites", *Aeronautical J.*, 368–372.

5. Broutman, L.J. (1967) "Fiber-reinforced plastics," in *Modern Composite Materials*, Editors, L.J. Broutman and H. Krock, Addison-Wesley, New York, Chapter 13.

6. Sanker, B. and Sun, C.T. (1986) "Low-velocity impact damage in graphite epoxy laminates subjected to tensile initial stresses," *AIAA J.*, **24**(3), 470–472.

7. Sjoblom, P. (1987) "Simple design approach against low velocity impact damage," *Proc. 32nd SAMPE Symposium*, Anaheim, CA, pp. 529–539.

8. Takeda, N., Sierakowski, R.L., and Malvern, L.E. (1981) "Transverse cracks in glass/epoxy cross ply laminates impacted by projectiles," *J. Mater. Sci.*, **16**, 2008–2011.

9. Sierakowski, R.L. and Takeda, N. (1981) "An investigation of in-plane failure mechanisms in impacted fiber reinforced plates," in *Composite Materials*, Editors, K. Kawata and T. Akasaka, Proceedings, First U.S.–Japan Conference, Tokyo, pp. 12–21.

10. Dorey, G. (1987) "Impact damage in composites-development, consequences and prevention," *Proceedings of the 6th International Conference on Composite Materials* combined with the *2nd European Conference on Composite Materials*, London, pp 3.1–3.26.

11. Evans, R.E. and Masters, J.E. (1987) *A New Generation of Epoxy Composites for Primary Structural Applications: Materials and Mechanics*, ASTM STP 937, pp 413–436.

12. Awerbuch, J., and Hahn, H.T. (1976) "Hard object impact damage of metal matrix composites," *J. Comp. Mater.*, **10**, 231–257.

13. Husman, G.E., Whitney, J.M., and Halpin, J.C. (1975) *Residual Strength Characterization of Laminated Composite Subjected to Impact Loading*, ASTM STP 568, pp. 92–113.

14. Caprino, G. (1983) "On the prediction of residual strength for notched laminates," *J. Mater. Sci.*, **18**, 2269–2273.

15. Avva, V.S., Vala, J.R., and Jeyaseelan, M. (1986) *Effect of Impact and Fatigue Loads on the Strength of Graphite/Epoxy Composites*, ASTM STP 893, pp. 196–206.

16. Lal, K.M. (1983) "Residual strength assessment of low velocity impact damage of graphite-epoxy laminates," *J. Reinforced Plastics and Composites*, **2**, 226–238.

17. Avery, J.G., Bradley, S.H., and King, K.M. (1981) *Fracture Control in Ballistic-Damaged Graphite/Epoxy Wing Structure*, ASTM STP 743, pp. 338–359.

18. Rhodes, M.D., Williams, J.G., and Starnes, J.H. (1981) "Low velocity impact damage in graphite-fiber reinforced epoxy laminates," *Polymer Composites*, **2**(1), 36–44.

19. Williams, J.G., Anderson, M.S., Rhodes, M.D., Starnes, Jr., J.H., and Stroud, W.J. (1979) *Recent Developments in the Design, Testing, and Impact-Damage Tolerance of Stiffened Composite Panels*, NASA TM 80077.

20. Avva, V.S. (1983) *Effect of Specimen Size on the Buckling Behavior of Laminated Composites Subjected to Low Velocity Impact*, ASTM STP 808, pp. 140–154.

21. Hirschbuehler, K.R. (1987) *A Comparison of Several Mechanical Tests Used to Evaluate the Toughness of Composites*, ASTM STP 937, pp. 61–73.

22. Leach, D.C., Curtis, D.C., and Tamblin, D.R. (1987) *Delamination Behavior of Carbon Fiber/Poly(etheretherketone) (PEEK) Composites*, ASTM STP 937, pp. 358–380.

23. Malvern, L.E., Sun, C.T., and Liu, D. (1989) *Delamination Damage in Central Impacts at Subperforation Speeds on Laminated Kevlar/Epoxy Plates*, ASTM STP 1012, pp. 387–405.

24. Cantwell, W.J. and Morton, J. (1984) "Low velocity impact damage in carbon fibre reinforced plastic laminates," *Proceedings of the 5th International Congress on Experimental Mechanics*, Montreal, Canada, June 10–15, pp. 314–319.

25. Rotem, A. (1988) "Residual flexural strength of FRP composite specimens subjected to transverse impact loading", *SAMPE J.*, **24**, 19–25.

26. Cantwell, W.J., Curtis, P., and Morton, J. (1984) "A Study of the Impact Resistance and Subsequent O-Compression Fatigue Performance of Non-Woven and Mixed Woven Composites," in *Structural Impact and Crashworthiness*, 2, Editor, J. Morton, Elsevier Applied Science Publ., London, pp. 521.

27. Llorente, S. (1989) "Damage tolerance of composite shear panels," *J. Amer. Helicopter Society*, **34**(2), 43–51.

28. Grady, J.E. and Meyn, E.H. (1989) "Vibration testing of impact-damaged composite laminates," *Proceedings 30th AIAA/ASME/ASC Conference*, Mobile, Alabama, April, pp. 2186–2193.

29. Duke, Jr., J.C. and Kiernan, M.T. (1988) "Impact damage development in damaged composite material," *Proceedings 4th U.S.–Japan Conference on Composite Materials*, Washington, D.C., June, pp. 63–71.

30. Moon, S.M., Hahn, H.T., and Jerina, K.L. (1988) "Detection of impact damage in composite laminates by acoustic-ultrasonic technique," *Proceedings 4th U.S.–Japan Conference on Composite Materials*, Washington, D.C., June, pp. 92–105.

31. Bishop, S.M. (1985) "The mechanical performance and impact behavior of carbon fiber reinforced PEEK," *Composite Structures*, **3**, 295–318.

32. Butcher, B.R. (1979) "The impact resistance of unidirectional CFRP under tensile stress," *Fiber Sci. Technol.*, **12**, 295–326.

33. Bowles, K.J. (1988) *The Correlation of Low Velocity Impact Resistance of Graphite Fiber Reinforced Composites with Matrix Properties*, ASTM STP 972, pp. 124–142.

34. Cantwell, W., Curtis, P., and Morton, J. (1983) "Post impact fatigue performance of carbon fibre laminates with non-woven and mixed woven layers," *Composites*, **14**(3), 301–305.

35. Cantwell, W.J., Curtis, P., and Morton, J. (1986) "An assessment of the impact performance of CFRP reinforced with high-strain carbon fibres," *Comp. Sci. Technol.*, **25**(2), 133–148.

36. Peiffer, D.G. (1979) "Impact strength of thick interlaminar composites," *J. Appl. Polym. Sci.*, **24**, 1451–1455.

37. Ross, C.A. and Sierakowski, R.L. (1973) Studies on the impact resistance of composite plates," *Composites*, **4**, 157–161.

38. Cantwell, W.J. (1988) "The influence of target geometry on the high velocity response of CFRP," *Composite Structures*, **10**(3), 247–265.

39. Malvern, L.E., Sun, C.T., and Liu, D. (1987) "Damage in composite laminates from central impacts at subperforation speeds," in *Recent Trends in Aeroelasticity, Structures and Dynamics*, Proceedings of the Symposium in Memory of Prof. Bisplinghoff, Editor, P. Hajela, University of Florida Press, Gainesville, FL, pp. 298–312.

40. Malvern, L.E. and Sun, C.T. (1986) *Delamination Sensing and Modeling in Localized Impacts on Filament-Reinforced Laminated Plates*, Final Report, US Army Research Office.

41. Cristescu, N., Malvern, L.E., and Sierakowski, R.L. (1975) *Failure Mechanisms in Composite Plates Impacted by Blunt-ended Penetrators*, ASTM STP 568, pp. 159–172.

42. Adsit, N.R. and Waszczak, J.P. (1979) *Effect of Near Visual Damage on the Properties of Graphite/Epoxy*, ASTM STP 674, pp. 101–117.

43. Avery, J.G. and Porter, T.R. (1975) *Comparison of the Ballistic Response of Metals and Composites for Military Applications*, ASTM STP 568, pp. 3–29.

44. Butcher, B.R. and Fernback, P.J. (1981) "Impact resistance of unidirectional CFRP under tensile stress: further experimental variables," *Fiber Sci. Technol.*, **14**, 41–58.

45. Cantwell, W.J. and Morton, J. (1985) "Detection of impact damage in CFRP laminates," *Composite Structures*, **3**, 241–257.

46. Clerico, M., Ruvinetti, G., Cipri, F., and Pelosi, M. (1989) "Analysis of impact damage and residual static strength in improved CFRP," *Int. J., Mat. and Product Tech.*, **4**(1), 61–70.

47. Curtis, P. and Bishop, S.M. (1984) "An assessment of the potential of woven carbon fibre-reinforced plastics for high performance applications," *Composites*, **15**(4), 259–264.

48. Demuts, E., Whitehead, R.S., and Deo, R.B. (1985) "Assessment of damage tolerance in composites," *Composite Structures*, **4**, 45–48.

49. Gandhe, G.V. and Griffin, Jr., O.H. (1989) "Post-impact characterization of interlined composite materials", *SAMPE Quarterly*, **20**, 55–58.

50. Griffin, C.F. (1987) *Damage Tolerance of Toughened Resin Graphite Composites*, ASTM STP 937, pp. 23–33.

51. Llorente, S. and Mar, J.W. (1989) "The residual strength of filament wound graphite/epoxy sandwich laminates due to impact damage and environmental conditioning," AAIA Paper 89-1275-CP, *Proceedings of the 30th AIAA/ASME/ASCE/AHS/ASC, Struct., Structures Dynamic Mater., Conf.,* Mobile, AL.

52. Ramkumar, R.L., Grimes, G.C., and King, S.J. (1986) *Characterization of T300/V-378A Graphite/Bismaleimide for Structural Applications*, ASTM STP 893, pp. 48–63.

53. Sharma, A.V. (1981) *Low Velocity Impact Tests on Fibrous Composite Sandwich Structures*, ASTM STP 734, pp 54–70.

54. Takamatsu, K., Kimura, J., and Tsuda, N. (1986) "Impact resistance of advanced composite structures," *Composites '86: Recent Advances in Japan and the United States*, Editors, K. Kawata, S. Umekawa, and A. Kobayashi, Proc. Japan–U.S. CCM-III, Tokyo, pp. 77–84.

55. Verpoest, I., Marien, J., Devos, J., and Wevers, M. (1987) "Absorbed energy, damage and residual strength after impact of glass-fiber epoxy composites", *Proceedings of the 6th International Conference on Composite Materials combined with the 2nd European Conference on Composite Materials*, London, England, pp. 3.485–3.489.

56. Williams, J.G. and Rhodes, M.D. (1982) *Effect of Resin on Impact Damage Tolerance of Graphite/Epoxy Laminates*, ASTM STP 787, pp. 450–480.

CHAPTER 6

IMPACT DAMAGE MODELING

6.1 INTRODUCTION

Perhaps the greatest attention to modeling impact damage in composites, whether by analytical or numerical techniques, has been focused at the subperforation impact level. Considerable attention has been focused on predicting the extent of damage as it may affect the static as well as the dynamic performance of the composite system subjected to impact damage. Therefore, such material characteristics as strength and stiffness have been studied extensively.

6.2 MODEL CLASSIFICATION

In general, the damage prediction models which have been studied for various impact velocity regimes can be classified into the following groups:

- empirical;
- semi-empirical;
- analytical.

The greatest focus of research interest has been in the first two of the above classification areas. Some typical models are presented here in order that the reader may get a flavor for the modeling approaches used and results obtained while some useful references which address the impact problem are included in [1]. For each of the models discussed, the classification issues raised earlier, that is, striker/target classification, target material, striker velocity, and other important parameters have been presented at the beginning of each problem considered. Introduction of a problem statement can help in the identification and categorization of sets of problems studied in the literature. This point can not be overemphasized since in studying the literature a systematic approach to categorizing the regime and the important parameters is necessary for understanding the parameters studied by each investigator. Therefore, at the beginning of each of the analytical examples cited, a suggested codification of the important parameters necessary to compare models studied is presented.

6.2.1 Empirical Modeling

A considerable amount of parametric testing of impacted composites has been performed in recent years in order to evaluate the damage response of composites as chronicled for example in reference [1]. In general damage results from such tests are highly dependent upon the striker velocity, type (nose shape), mass, as well as target material, and target thickness. One such empirical data base for evaluating the residual strength of composite targets is due to Avery and Porter [2]. A codification of the data base as identified by these investigators is given in Table 6.1.

The approach adopted by Avery and Porter is to test both metal and composite structures subjected to ballistic impact at perforation speeds for a variety of strikers, angles of obliqueness and striker velocities. Parameters investigated include target thickness, striker velocity, impact angle, and striker type. The damage model introduced is defined in terms of an observed transverse lateral damage (TLD) which in turn is taken as a measure of a critical flaw size and used as a correlating measure to a critical stress intensity factor. This can be stated in a equation form as

$$\lambda_c = \sigma_c \left[\frac{\pi(\text{TLD})_{\text{eff}}}{2} \right]^{1/2}$$

TABLE 6.1 Problem Statement (Avery and Porter [2])

Striker material	Steel
Striker geometry	0.30 caliber (armor piercing projectile)
	0.30 caliber ball
	0.50 caliber (armor piercing projectile)
Striker velocity range	1000 to 3000 ft/s
Striker classification	Non-deformable
Target material	6061-T6 aluminum alloy,
	7075-T6 aluminum alloy
	Thornel (50S) graphite-epoxy (5206)
	Boron-epoxy
Target geometry	18 in × 36 in × 0.190 in
Target velocity	0
Target classification	Deformable
Test procedure	Ballistic
Damage modes	Perforation
Damage modeling	"Effective" critical
	Stress intensity factor
Damage observation	Visual

where σ_c = static residual fracture stress after ballistic damage. $(TLD)_{eff}$ = effective damage size.

The above result is similar to the equation for Griffith's stress intensity factor as derived from classical elasticity theory. For composites and for impacts occurring at an angle of obliqueness, an effective (TLD) is suggested as a measure of the lateral damage. Such a measure is taken in order that the observed damage should be based upon a projected striker diameter. Results from this test technique are graphically shown in Figure 6.1.

The above figure relates fracture strength with damage size to different types of strikers fired at several angles of obliqueness and velocities. This data base can then be used as a measure to establish design guidelines for impact problems, that is those related to structural integrity requirements.

It should be noted that in these tests the data obtained were collected for relatively thin composites determined over a selected range of impact conditions. It is further observed that when impact tests were conducted with an applied tensile pre-load, the stress necessary to fracture the tested specimens was reduced significantly relative to the unloaded specimen case. It is also to be noted that laboratory

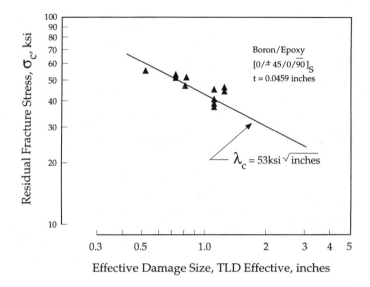

Figure 6.1   Residual fracture stress as a function of effectivie damage size of projectile damaged boron-epoxy panels [2]. (Copyright ASTM. Reprinted with permission.)

tests of the type studied, by the investigators, do not generally match results obtained involving real structural configurations, which also include stiffening elements.

6.2.2   Semi-Empirical Modeling

As an example of a semi-empirical scheme used for evaluating the residual strength of composite systems, the approach of Husman et al. [3] is cited. For this example, the information is as presented in Table 6.2.

The principal information obtained using this approach is that the residual strength of impacted specimens can be obtained by performing two experiments: a tensile test performed on an unflawed test coupon and a separate tensile test performed on an impacted test coupon. The model is based upon the idea that the damage indicated on the coupon during single-point impact can be modeled by an equivalent crack size which can be determined and correlated with the kinetic impact energy and residual strength of the impacted coupon, using a classical linear fracture mechanics approach. This model has been discussed in detail in Chapter 5 of this book.

TABLE 6.2 Problem Statement (Husman et al. [3])

Striker material	Steel
Striker geometry	0.177 in and 0.25 in diameter spherical projectiles
Striker velocity range	100–600 ft/s
Striker classification	Non-deformable
Target material	E glass-epoxy and graphite-epoxy composite (0°, 90°) symmetric laminates
Target geometry	6.0 in × 0.5 in. straight sides tensile coupons with thickness 0.1 in to 0.14 in
Target classification	Deformable
Test procedure	Ballistic impact
Damage modes	Matrix cracking, delamination, and perforation
Damage modeling	Energy balance
Damage observation	Visual

A similar approach to predict the damage zone size and residual strength in metal-matrix composites has been used by Awerbuch and Hahn [4]. In this study the ballistic tests were carried out on boron/ aluminum and borsic/titanium composite cantilever plates using 0.177 in diameter steel sphere strikers propelled over a velocity range of 50–4000 ft/s. The analytical model proposed for explaining the data obtained has been based upon classical fracture mechanics.

Impact Force and Duration   The previous examples have focused on striker impact velocities sufficient to perforate the target. A number of important practical problems, however, involve the low-velocity impact of composites, such as when a tool drops on a structural composite member. The result of this impact is to produce damage within the composite member. This damage may be invisible to the naked eyes but it may degrade significantly the mechanical properties of the composite, in particular, the stiffness and strength of the structural element. In this problem type, the indentation of the striker into the target at the impact point must be considered in addition to the global deformations of the target. To study the local deformation effect, indentation using Hertzian contact mechanics has been used. Solutions generated by this approach allow for the prediction of stress and deformation near the local contact point as a function of the geometrical and material properties of the striker and target. One

TABLE 6.3 Problem Statement (after Skivakumar et al. [5])

Striker material	Steel
Striker geometry	19 mm radius sphere
Striker velocity range	0–100 ft/s
Striker classification	Non-deformable
Target material	Steel, aluminum, graphite-epoxy, quasi-isotropic laminated composite
Target geometry	25.4 mm radius, 1.04 mm thickness circular plates
Target velocity	0
Target classification	Deformable
Test procedure	Drop test
Damage modes	Displacement/deformation, perforation
Damage modeling	Energy balance and spring–mass models
Damage observation	Visual, X-ray

such analysis using this approach is that by Shivakumar et al. [5]. The problem statement is given in Table 6.3.

Two models, an energy-balance model and a spring–mass model, have been introduced to determine information on the impact force and its duration on the target. To determine the force acting on the target structure, an energy-balance model relating the kinetic energy of the striker to the energy due to contact, bending, transverse shear, and membrane deformations have been considered. This approach includes deformations due to Hertzian contact previously studied by Greszczuk [6] as well as the Hertzian contact and plate flexural deflections as studied for example, by Sun and Chattopadhyay [7] and Dobyns [8]. As presented in the example here, it is assumed that the dominant mode of vibration of the plate is the first mode and that the effects of higher modes are small and can be neglected. The following two models for the problem stated above can be useful:

1. energy-balance model;
2. spring–mass model.

Energy-Balance Model The geometrical configuration adopted for study is that of a circular composite material plate target of radius r_0 and thickness h impacted by a rigid spherical striker of a radius r_i, mass m_i, traveling at a velocity of v_i (Figure 6.2). The composite

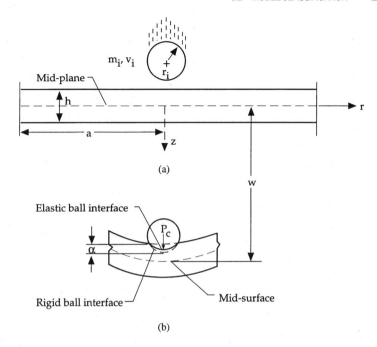

Figure 6.2 Central transverse impact on a circular plate [5]: (a) before impact $(t < 0)$; (b) during impact $(t > 0)$. (Reprinted with permission from ASME.)

laminate is considered to be quasi-isotropic in construction. Upon impact, the target undergoes deformations consisting of indentation, bending, transverse shear, and membrane stretching. The membrane deformation becomes important when the displacement/thickness ratio ≥ 0.2, otherwise, its contribution towards the total energy of deformation can be neglected.

The energy-balance model can be expressed analytically by relating the maximum kinetic energy of the striker at time $t = 0$ to the aforementioned deformations. That is

$$\tfrac{1}{2}m_i v_i^2 = E_c + E_{bs} + E_m \tag{6.1}$$

where E_c, E_{bs}, and E_m are the contact, bending-shear, and membrane energies, respectively. Energy losses associated with material damping, surface friction, and higher modes of vibration are neglected. The energies E_c, E_{bs}, and E_m can be calculated as follows. The energy E_c, due to indentation or contact deformation, depends upon the form of the contact force P_c and the contact deformation α (see Figure 6.2), calculated as

$$E_c = \int_0^\alpha P_c \, d\alpha \tag{6.2}$$

The impact force, P_c, varies with the contact deformation, and can be correlated with α through the Hertz law of contact [9] as

$$P_c = \eta \alpha^{3/2} \tag{6.3}$$

The parameter η is known as the contact stiffness parameter and depends upon the material, geometrical, and surface properties of both the target and the striker. Various research efforts have been made on obtaining appropriate values for the parameter η. For example, η for a spherical isotropic rigid striker intersecting with a transversely isotropic composite plate can be given as [10]

$$\eta = \frac{4\sqrt{r_i}}{3\pi(k_1 + k_2)} \tag{6.4}$$

where

$$k_1 = \frac{(1 - \nu_i^2)}{\pi E_i}$$

and

$$k_2 = \frac{\sqrt{A_{22}}[(\sqrt{A_{11}A_{22}} + G_{zr})^2 - (A_{12} + G_{zr})^2]^{1/2}}{2\pi\sqrt{G_{zr}}(A_{11}A_{22} - A_{12}^2)}$$

$$A_{11} = E_z(1 - \nu_r)\beta$$

$$A_{22} = \frac{E_r\beta(1 - \nu_{zr}^2\delta)}{(1 + \nu_r)}$$

$$A_{12} = E_r\nu_{zr}\beta$$

$$\beta = \frac{1}{1 - \nu_r - 2\nu_{zr}^2\delta}$$

$$\delta = E_r/E_z$$

where

E_i, ν_i = Young's modulus and Poisson's ratio of the impactor, respectively

E, G, ν = Young's modulus, shear modulus, and Poisson's ratio, respectively, of the plate, with subscripts r and z, referring to the radial and thickness directions, respectively.

By combining Eqs. (6.2) and (6.3) and simplifying, the contact energy, E_c, can be obtained as

$$E_c = \frac{2}{5}\left(\frac{P_c^{5/3}}{\eta^{2/3}}\right) \tag{6.5}$$

The reactive forces due to bending, shear, and membrane deformations, are also obtained using force-deformation relations reported by Volmir [11]:

The bending-shear force, $P_{bs} = k_{bs}w$

The membrane force, $P_m = k_m w^3$ $\qquad$ (6.6)

where

$$k_{bs} = \frac{k_b k_s}{k_b + k_s}, \text{ the effective stiffness}$$

k_b, k_s, k_m = bending, shear, and membrane stiffness, respectively, of the target plate

The shear stiffness, k_s, for transversely loaded circular laminates has been determined by Lukasiewicz [12] as

$$k_s = \frac{4\pi G_{zr} h}{3}\left[\frac{E_r}{E_r - 4\nu_{rz}G_{zr}}\right]\left[\frac{1}{\frac{4}{3} + \log\,(a/a_c)}\right] \tag{6.7}$$

where the contact radius, a_c, depends upon the total force of both the bending-shear and membrane effects, as given by [6]

$$a_c = \left[\frac{3\pi}{4}(P_{bs} + P_m)(k_1 + k_2)r_i\right]^{1/3} \tag{6.8}$$

Since the value of $(P_{bs} + P_m)$ is initially unknown, this can be estimated using the contact radius value as half of the target thickness.

Based on the above considerations, the bending-shear energy and membrane energy can easily be obtained by integrating Eq. (6.6) with respect to the displacement variable (w). The energy balance Eq. (6.1) can then be written as

$$m_i v_i^2 = k_{bs} w^2 + \frac{k_m w^4}{2} + \frac{4}{5} \left[\frac{(k_{bs} w + k_m w^3)^5}{\eta^2} \right]^{1/3} \qquad (6.9)$$

For a given m_i and v_i, the deflection w can easily be calculated from the above equation, and in turn used to obtain the total impact force for a specified striker-target system. Some typical results for an impacted clamped quasi-isoptropic composite plate having the properties given in Table 6.4 are shown in Figure 6.3.

The thicknesses selected represent a moderately thick $(h = 3.2$ mm), intermediate $(h = 1.6$ mm), and a thin target $(h = 0.8$ mm). The membrane effect is shown to be large for the thin plate, while transverse shear becomes significant for moderately thick plates. At high impact velocities, and as the target deflection increases, the membrane stiffening effect is found to alleviate shear flexibility.

Spring-Mass Model   To predict the time-history of the impact force, spring-mass models are useful. One such simple model is discussed below [5]. The model is based upon the following key assumptions:

1. The striker and the target plate can be represented by masses m_i and m_p, respectively. The effective plate mass is included as one quarter of the total plate mass.
2. The transverse load-deformation behavior of the plate is represented by combined bending, shear, and membrane springs.

TABLE 6.4   Properties of an Impacted Clamped Plate

Material	E_r (GPa)	E_z (GPa)	G_r (GPa)	G_z (GPa)	ν_r	ν_{zr}	ρ (density) (kg/m^3)
Graphite-epoxy (T300/5208)	50.81	11.78	19.38	4.11	0.31	0.06	1611

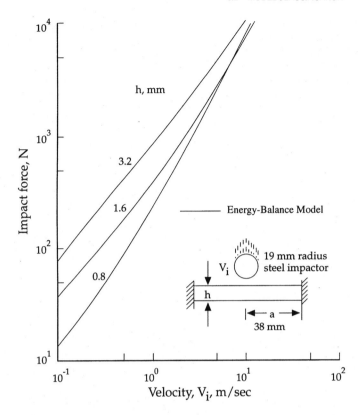

Figure 6.3 Predicted impact force for clamped (immovable support) composite plates (graphite-epoxy quasi-isotropic laminate) [5]. (Reprinted with permission from ASME.)

3. For thin plates the bending-shear stiffness $k_{bs} \approx k_b$, bending stiffness.
4. For small plate deflections ($w < 0.2h$), the membrane-stretching force is assumed to be $P = K_m w^3$.
5. Plate damage, surface friction and material dumping are neglected.

The spring–mass model for the case of two degrees of freedom is shown in Figure 6.4 with plate bending and shear stiffnesses for four types of plate boundaries presented in Table 6.5.

The displacements $x_1(t)$ and $x_2(t)$ are taken as the responses of masses m_i and m_p at any time t after impact, that is when the striker

TABLE 6.5 Bending and Membrane Stiffness Parameters of Centrally Loaded Plates [5][*]

Boundary Condition	Edge Conditions	Bending Stiffness (K_b)	Membrane Stiffness Parameters (K_m)
Clamped	Immovable	$\dfrac{4\pi E_r h^3}{3(1 - \nu_r^2)a^2}$	$\dfrac{(353 - 191\nu_r)\pi E_r h}{648(1 - \nu_r)a^2}$
	Movable	$\dfrac{4\pi E_r h_3}{3(1 - \nu_r^2)a^2}$	$\dfrac{191\pi E_r h}{648a^2}$
Simply supported	Immovable	$\dfrac{4\pi E_r h^3}{3(3 + \nu_r)(1 - \nu_r)a^2}$	$\dfrac{\pi E_r h^3}{(3 + \nu_r)^4 a^2} \left\{ K^* + \dfrac{1}{(1 - \nu_r)} \left[\dfrac{(1 + \nu_r)^4}{4} \right. \right.$ $\left. \left. +2(1 + \nu_r)^3 + 8(1 + \nu_r)^2 + \right. \right.$ $\left. \left. +16(1 + \nu_r) + 16 \right] \right\}$
	Movable	$\dfrac{4\pi E_r h^3}{3(3 + \nu_r)(1 - \nu_r)a^2}$	$\dfrac{\pi E_r h}{a^2(3 + \nu_r)^4}[K^*]$

$$K^* = \tfrac{191}{648}(1 + \nu_r)^4 + \tfrac{41}{27}(1 + \nu_r)^3 + \tfrac{32}{9}(1 + \nu_r)^2 + \tfrac{40}{9}(1 + \nu_r) + \tfrac{8}{3}]$$

[*] Reprinted with permission from ASME.

and target are assumed to be in constant contact with each other. If the striker is taken as being rigid and the target deflection as $x_2(t) = w$, then the contact deformation can be written as $\alpha = x_1(t) - x_2(t)$.

Equilibrium equations for the striker–target configuration (Figure 6.4) are given as

$$m_i \ddot{x}_i + \lambda n[x_1 - x_2]^{1.5} = 0 \qquad (6.10)$$

$$m_p \ddot{x}_2 + k_{bs}x_2 + k_m x_2^3 - \lambda n[x_1 - x_2]^{1.5} = 0 \quad (6.11)$$

with the coefficient

$$\lambda = 1 \quad \text{for} \quad x_1 > x_2,$$
$$\lambda = -1 \quad \text{for} \quad x_1 < x_2$$

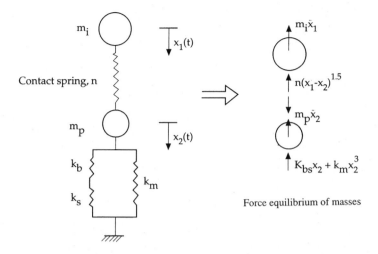

Figure 6.4   Two degree-of-freedom spring–mass (TDOF S-M) model for low-velocity impact of a circular plate [5]. (Reprinted with permission from ASME.)

and where $\ddot{x}_1$, $\ddot{x}_2$ = accelerations d^2x_1/dt^2 and d^2x_2/dt^2, respectively. The initial conditions for the striker and target are

$$m_i : \qquad x_1(0) = 0, \quad \dot{x}_1(0) = v_i \qquad\qquad (6.12)$$
$$m_p : \qquad x_2(0) = 0, \quad \dot{x}_2(0) = 0 \qquad\qquad (6.13)$$

A numerical scheme is used to solve the nonlinear coupled differential equations, with the solution to be terminated when $x_2(t)$ becomes zero or reaches a negative value. In order to find the impact force-history, the calculated values of $x_2(t) = w(t)$ as a function of time t are used in Eq. (6.6).

In some instances the coupled system of differential equations given by Eqs. (6.10) and (6.11) can be simplified when m_i is greater than 3.5 times the total plate mass. The governing equation then represents a single degree of freedom system in which $x_1 = (w + \alpha)$, and the initial conditions become at $t = 0$, $x_1(0) = 0$, $\dot{x}_1(0) = v_i$. By obtaining the solution for $x_1(t)$, the impact force history $p(t)$ can be determined as noted above.

Some typical results are shown in Figures 6.5 and 6.6, for the case of an eight-ply graphite-epoxy composite target plate of 45 mm

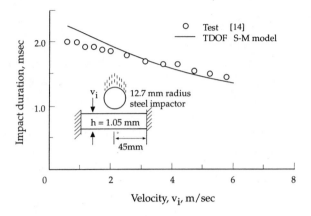

Figure 6.5 Predicted by two degree-of-freedom spring–mass model or TDOF S-M and measured impact duration for a clamped graphite-epoxy quasi-isostropic laminate [5]. (Reprinted with permission from ASME.)

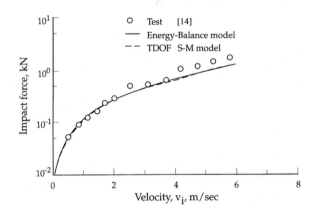

Figure 6.6 predicted by TDOF S-M and energy balance (E–B) models and measured impact force for a clamped graphite/epoxy quasi-isotropic laminate [5]. (Reprinted with permission from ASME.)

radius impacted by a rigid steel sphere at several impact velocities. Both the impact force duration and magnitude agree well with the two degrees of freedom model discussed above. Only at low velocities does the impact force duration deviate from agreement with that of the model predictions.

6.2.3 Analytical Models of Damage

The preceding discussion describes some of the difficulties in defining the details of the impact event. These difficulties present impediments to the development of predictive methodologies to define impact damage caused by applied energy levels which in turn can be used to assess what effect the induced damage has on the material/structure residual strength. The effect of damage on the residual strength can be determined through testing; however, such a technique is both time consuming and expensive. Thus, there is a need to predict the nature of the damaged region for defined impact energy levels in composite structures and for specified material properties and structural configurations. In general, beam and plate models are commonly used as structural elements for evaluating composite damage due to impact. However, in the determination of damage development in composites, the following issues need to be addressed:

1. determining the load-time history of the impact force;
2. determining the stress-time history of the target internal stresses;
3. determination of failure modes in the target.

One of the important failure modes observed in laminated composites subjected to impact loads is that of delamination which is governed by transverse shear and transverse normal stresses ocurring at the interlaminar planes of a laminate. The prediction of delamination requires a detailed knowledge of the transverse shear and normal stresses. One such early analysis has been studied by Dobyns [8]. In this analysis, a simply supported plate subjected to a dynamic loading distributed over a small contact area has been studied. The contact area over which the force is distributed has been selected to be of the order of the projected cross sectional area of the striker. With this approach, the transverse shear force remains finite over the contact area. Delaminations in a given composite plate are assumed to take place when the transverse shear force exceeds a threshold value. A solution by Dobyns [8] to the plate-impact problem follows. The assumed displacement field, including transverse shear effects is given by

$$u = u^0(x, y, t) + z\psi_x(x, y, t)$$
$$v = v^0(x, y, t) + z\psi_y(x, y, t)$$
$$w = w(x, y, t), \tag{6.14}$$

The quantities u^0, v^0 and w represent the displacements at the midplane of the plate in the x, y, and z directions, respectively. The quantities ψ_x and ψ_y represent the shear rotations in the x and y directions, respectively. For analysis, the following assumptions are made:

- The plate is specially orthotropic: $B_{ij} = 0$, $A_{16} = A_{26} = D_{16} = D_{26} = 0$.
- A uniform initial stress, that is, N_x^0, N_y^0 is assumed to exist.
- A foundation stiffness K is included.

The equations of motion are then given by Whitney and Pagano [13] as

$$D_{11}\psi_{x,xx} + D_{66}\psi_{x,yy} + (D_{12} + D_{66})\psi_{y,xy}$$
$$- kA_{55}\psi_x - kA_{55}w_{,x} + m_x = I\ddot{\psi}_x \tag{6.15a}$$
$$(D_{12} + D_{66})\psi_{x,xy} + D_{66}\psi_{y,xx} + D_{22}\psi_{y,yy}$$
$$- kA_{44}\psi_y - kA_{44}w_{,y} + m_y = I\ddot{\psi}_y \tag{6.15b}$$
$$kA_{55}\psi_{x,x} + (kA_{55} + N_x^0)w_{,xx} + kA_{44}\psi_{y,y} + (kA_{44} + N_y^0)w_{,yy}$$
$$+ P_z + Kw = P\ddot{w} \tag{6.15c}$$

The comma denotes differentiation with respect to x and y, while the dot indicates differentiation with respect to time t. The quantities P_z, m_x, m_y represent the distributed loads, k is the Mindlin shear correction factor, and the stiffness and inertias are given by

$$(A_{ij}, D_{ij}) = \int_{-h/2}^{h/2} Q_{ij}(I, z^2)\, dz, \qquad i, j = 1, 2, 6$$

$$A_{ij} = \int_{-h/2}^{h/2} C_{ij}\, dz, \qquad\qquad i, j = 4, 5 \tag{6.16}$$

$$(P, I) = \int_{h/2}^{h/2} \rho(1, z^2)\, dz$$

The Q_{ij} in Eq. (6.16) represent the reduced in-plane stiffnesses while the C_{ij} are the transverse shear stiffnesses. The plate considered in this analysis is a simply supported, rectangular plate, with dimensions $(a \times b)$ and uniform thickness.

Boundary conditions are given by

$$
\begin{aligned}
w = \psi_{x,x} = 0 \qquad & \text{at } x = 0, a \\
w = \psi_{y,y} = 0 \qquad & \text{at } y = 0, b
\end{aligned}
\tag{6.17}
$$

In the dynamic loading case, the frequency of natural vibration can be computed from the assumed displacements, that is,

$$
\psi_x = \psi_x e^{i\omega t}, \quad \psi_y = \psi_y e^{i\omega t}, \quad w = W e^{i\omega t}
\tag{6.18}
$$

where ψ_x, ψ_y and W are given by

$$
\begin{aligned}
\psi_x &= A_{mn} \cos (m\pi x/a) \sin (n\pi y/b) \\
\psi_y &= B_{mn} \sin (m\pi x/a) \cos (n\pi y/b) \\
W &= C_{mn} \sin (m\pi x/a) \sin (n\pi y/b)
\end{aligned}
\tag{6.19}
$$

substituting (6.18) into (6.15) gives

$$
D_{11}\psi_{x,xx_{mn}} + D_{66}\psi_{x,yy_{mn}} + (D_{12} + D_{66})\psi_{y,xy_{mn}}
$$

$$
- kA_{55}\psi_{x_{mn}} - kA_{55}W_{,x_{mn}} = -\omega_{mn}^2 I \psi_{x_{mn}}
\tag{6.20a}
$$

$$
(D_{12} + D_{66})\psi_{x,xy_{mn}} + D_{66}\psi_{y,xx_{mn}} + D_{22}\psi_{y,yy_{mn}}
$$

$$
- kA_{44}\psi_{y_{mn}} - kA_{44}W_{,y_{mn}} = -\omega_{mn}^2 I \psi_{y_{mn}}
\tag{6.20b}
$$

$$
kA_{55}\psi_{x,x_{mn}} + (kA_{55} + N_x^0)W_{,xx_{mn}} + kA_{44}\psi_{y,y_{mn}} + (kA_{44} + N_y^0)W_{,yy_{mn}}
$$

$$
+ KW_{mn} = -\omega_{mn}^2 P W_{mn}
\tag{6.20c}
$$

The natural frequencies can be found by substituting (6.19) into (6.20a), (6.20b), and (6.20c), giving

$$
\begin{bmatrix} L'_{11} L_{12} L_{13} \\ L_{12} L'_{22} L_{23} \\ L_{13} L_{23} L'_{33} \end{bmatrix}
\begin{Bmatrix} A'_{mm} \\ B'_{mn} \\ C'_{MN} \end{Bmatrix}
=
\begin{Bmatrix} 0 \\ 0 \\ 0 \end{Bmatrix}
\tag{6.21}
$$

where

$$L_{11} = D_{11}(m\pi/a)^2 + D_{66}(n\pi/b)^2 + kA_{55}$$
$$L_{12} = (D_{12} + D_{66})(m\pi/a)(n\pi/b)$$
$$L_{13} = kA_{55}(m\pi/a)$$
$$L_{22} = D_{66}(m\pi/a)^2 + D_{22}(n\pi/b)^2 + kA_{44}$$
$$L_{23} = kA_{44}(n\pi/b)$$
$$L_{33} = (kA_{55} + N_x^0)(m\pi/a)^2 + (kA_{44} + N_y^0)(n\pi/b)^2 + K$$
$$L_{11}' = L_{11} - \omega_{mn}^2 I, \; L_{22}' = L_{22} - \omega_{mn}^2 I, \; L_{33}' = L_{33} - \omega_{mn}^2 P$$

For each m and n pair, three eigenvalues and eigenvectors exist. If the rotational inertia term is suppressed, then for each m and n only one eigenvalue and eigenvector results:

$$\omega_{mn}^2 = (QL_{33} + 2L_{12}L_{23}L_{13} - L_{22}L_{13}^2 - L_{11}L_{23}^2)/(PQ)$$
$$Q = L_{11}L_{22} - L_{12}^2 \tag{6.22}$$

The eigenvectors associated with the natural frequencies normalized with respect to C_{mn}' are given by

$$A_{mn}' = \frac{L_{12}L_{23} - L_{22}L_{13}}{L_{11}L_{22} - L_{12}^2} C_{mn}', \quad B_{mn}' = \frac{L_{12}L_{13} - L_{11}L_{23}}{L_{11}L_{22} - L_{12}^2} C_{mn}' \tag{6.23}$$

Also, the orthogonality condition for the principal modes is given by

$$(\omega_{mn}^2 - \omega_{pq}^2) \int_0^a \int_0^b (PW_{mn}W_{pq} + I\psi_{x_{mn}}\psi_{x_{pq}} + I\psi_{y_{mn}}\psi_{y_{pq}}) \, dx \, dy = 0 \tag{6.24}$$

The integral in (6.24) is zero for $m, n \neq p, q$. Separating the equations of motion into a position- and time-dependent solution given by

$$\psi_x(x, y, t) = \sum_m \sum_n \psi_{x_{mn}}(x, y)T_{mn}(t)$$

$$\psi_y(x, y, t) = \sum_m \sum_n \psi_{y_{mn}}(x, y)T_{mn}(t) \tag{6.25}$$

$$w(x, y, t) = \sum_m \sum_n W_{mn}(x, y)T_{mn}(t)$$

and then substituting (6.25) into (6.15) gives

$$-\sum_m \sum_n \omega_{mn}^2 \psi_{x_{mn}} T_{mn} + \frac{m_x}{I} = \sum_m \sum_n \psi_{x_{mn}} \ddot{T}_{mn}$$

$$-\sum_m \sum_n \omega_{mn}^2 \psi_{y_{mn}} T_{mn} + \frac{m_y}{I} = \sum_m \sum_n \psi_{y_{mn}} \ddot{T}_{mn} \qquad (6.26)$$

$$-\sum_m \sum_n \omega_{mn}^2 W_{mn} T_{mn} + \frac{p_z}{P} = \sum_m \sum_n W_{mn} \ddot{T}_{mn}$$

The distributed loads m_x, m_y, and p_z can also be expanded in terms of the generalizerd $Q_{mn}(t)$ as

$$\frac{m_x}{I} = \sum_m \sum_n Q_{mn}(t) \psi_{x_{mn}}(x, y)$$

$$\frac{m_y}{I} = \sum_m \sum_n Q_{mn}(t) \psi_{y_{mn}}(x, y) \qquad (6.27)$$

$$\frac{p_z}{P} = \sum_m \sum_n Q_{mn}(t) W_{mn}(x, y)$$

The $Q_{mn}(t)$ can then be solved by multiplying the first equation by $I\psi_x$, the second by $I\psi_y$, and the third by PW. Adding these three equations and integrating over the plate area, taking into account orthogonality, gives

$$Q_{mn}(t) = \frac{\int_0^a \int_0^b (m_x \psi_{x_{mn}} + m_y \psi_{y_{mn}} + p_z W_{mn})\, dx\, dy}{\int_0^a \int_0^b (\psi_{x_{mn}}^2 I + \psi_{y_{mn}}^2 I + W_{mn}^2 P)\, dx\, dy} \qquad (6.28)$$

Substituting Eq. (6.27) into Eq. (6.26) gives for m, n,

$$\ddot{T}_{mn}(t) + \omega_{mn}^2 T_{mn}(t) = Q_{mn}(t) \qquad (6.29)$$

A solution to Eq. (6.29) for zero initial velocity and displacement is

$$T_{mn}(t) = \frac{1}{\omega_{mn}} \int_0^t Q_{mn}(\tau) \sin \omega_{mn}(t - \tau)\, d\tau \qquad (6.30)$$

The corresponding solutions for W, ψ_x, ψ_y are obtained by substituting Eqs. (6.19), (6.23), (6.28), and (6.30) into Eq. (6.25). Simplifications can be obtained for the cases in which the rotatory inertia terms can be suppressed. Thus, for $m_x = m_y = I = 0$, the generalized force Q_{mn} can be expressed as

$$Q_{mn}(t) = \frac{4}{PabC'_{mn}} \int_0^a \int_0^b p_z(x, y) \sin \frac{m\pi x}{a} \sin \frac{n\pi y}{b} \, dx \, dy, \quad (6.31)$$

The resultant expression for w, ψ_x, ψ_y using Eqs. (6.18), (6.23), (6.25), (6.28), and (6.30) are given as

$$w(x, y, t) = \frac{1}{P} \sum_m \sum_n \frac{q_{mn} \sin(m\pi x/a) \sin(n\pi y/b)}{\omega_{mn}}$$

$$\times \int_0^t F(\tau) \sin \omega_{mn}(t - \tau) \, d\tau \qquad (6.32a)$$

$$\psi_x(x, y, t) = \frac{1}{P} \sum_m \sum_n \left(\frac{L_{12}L_{23} - L_{22}L_{13}}{L_{11}L_{22} - L_{12}^2} \right)$$

$$\times \frac{q_{mn} \cos(m\pi x/a) \sin(n\pi y/b)}{\omega_{mn}} \int_0^t F(\tau) \sin \omega_{mn}(t - \tau) \, d\tau \quad (6.32b)$$

$$\psi_y(x, y, t) = \frac{1}{P} \sum_m \sum_n \left(\frac{L_{12}L_{13} - L_{11}L_{23}}{L_{11}L_{22} - L_{12}^2} \right)$$

$$\times \frac{q_{mn} \sin(m\pi x/a) \cos(n\pi y/b)}{\omega_{mn}} \int_0^t F(\tau) \sin \omega_{mn}(t - \tau) \, d\tau \quad (6.32c)$$

where $Q_{mn}(t) = (q_{mn}/PC'_{mn})$ and q_{mn} represent load distribution on the plate.

The convolution integrals expressed in Eq. (6.32) can be solved for any number of loading functions as shown in Figure 6.7.

For the case of an impact-loaded rectangular orthotropic panel, the response can be computed from Eq. (6.16) by considering the result for the rigid impact between a mass and a structural element as [8]

$$v_i t - (1/m) \int_0^t dt \int_0^t F \, dt = w_1(c) \qquad (6.33)$$

where v_i is the initial velocity of the striker of mass m; $w_1(c)$ is the transient response of the structure at the impacted point c as a func-

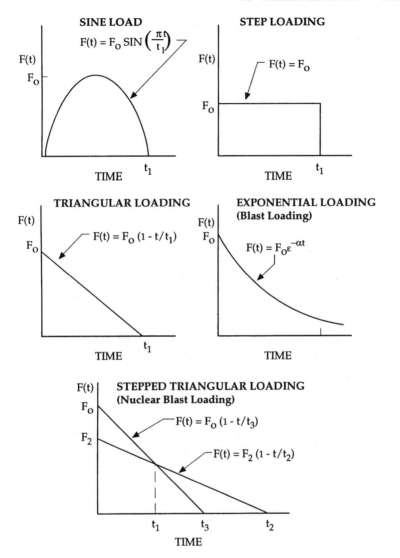

Figure 6.7 A variety of load pulse shapes [8]. (Copyright © 1981 AIAA. Reprinted with permission.)

tion of time; F is the contact force. For a non-rigid striker, Eq. (6.33) takes the following form [9]:

$$\alpha = w_2 - w_1(c) = v_i t - (1/m) \int_0^t dt \int_0^t F \, dt - w_1(c) \qquad (6.34)$$

where α is the difference between the displacement of the striker and deflection of the structure at the contact point $w_1(c)$ as shown in Figure 6.8.

Hertz's contact law can be used to relate the contact force and the deformation as

$$F = k_2 \alpha^{3/2} \qquad (6.35)$$

Substituting Eq. (6.35) into Eq. (6.34) gives an expression for the impact response of a plate as

$$\left(\frac{F}{k_2}\right)^{2/3} = v_i t - (1/m) \int_0^t dt \int_0^t F \, dt - w_1(c) \qquad (6.36)$$

where

$$k_2 = \frac{4\sqrt{r_i}}{3\pi(\delta_1 + \delta_2)} \qquad (6.37)$$

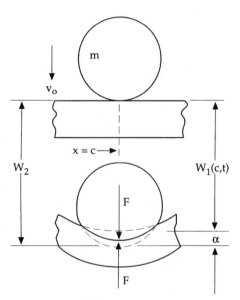

Figure 6.8 Central transverse impact of a rigid body with a contact surface [8]. (Copyright © 1981 AIAA. Reprinted with permission.)

and

$$\delta_1 = (1 - \nu_1^2)/E_1\pi$$
$$\delta_2 = (1 - \nu_2^2)/E_2\pi$$

where

r_i = striker radius

E_1, E_2 = Elastic moduli of the striker and plate, respectively

ν_1, ν_2 = Poisson's ratios of the striker and plate, respectively

Equation (6.36) is a nonlinear expression which requires a numerical solution. In order to obtain a finite normal shear force, the impact load is idealized as a uniform load over a very small rectangular area. Alternatively, the load can be considered as a cosine-shaped load over a small rectangular area. Some results obtained for the interlaminar shear stress, maximum bending strain, and maximum impact force on graphite-epoxy laminates of varying thicknesses are shown in Figure 6.9. The plate thickness is seen to have only a small effect on interlaminar shear stress but a larger effect on bending strain and impact force. Results obtained for the case of a low impact velocity, and a high impact velocity, for the same composite laminate, are shown in Figure 6.10. For a low-velocity event, the deflection shape is similar to that obtained for the case of static response. For large impact velocities panel deformation is limited to the impact regions such that panel size and boundary conditions are of less importance as in the case of the low-velocity impact event.

The effect of impact energy versus delamination for target composite plates has been studied for the case of both fabric and pre-preg graphite/epoxy composites as well as for different thicknesses and boundary conditions. Over the range of impact energies studied there is no visible delamination at the lowest impact energies; however, large delamination is evident for higher impact energy levels (Figure 6.11).

Further pertinent efforts for modeling deformation and damage modes in laminated composites include those of Choi and Chang [15] and Goldsmith et al. [16]. The approach developed by Choi and Chang aims at predicting the initiation of matrix cracking and delamination as well as the final damage state as a function of

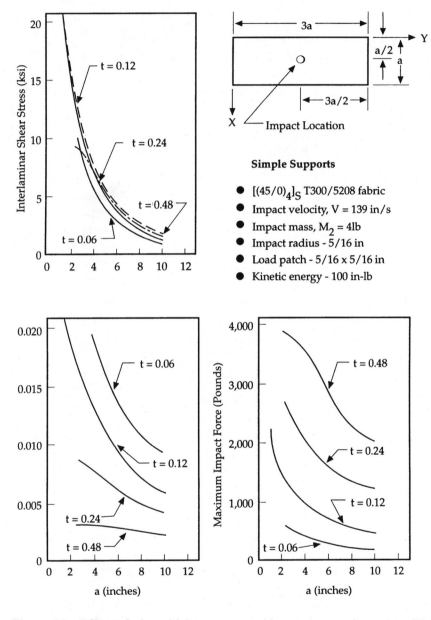

Figure 6.9 Effect of plate thickness on graphite-epoxy panel response [8]. (Copyright © 1981 AIAA. Reprinted with permission.)

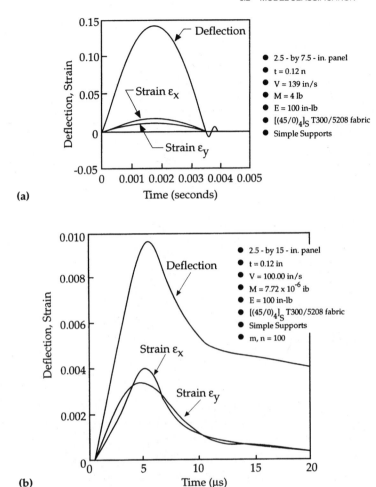

Figure 6.10 Deflection and strain vs. time for (a) low-velocity impact and (b) high-velocity impact [8]. (Copyright © 1981 AIAA. Reprinted with permission.)

impactor mass, laminate properties, and stacking sequence in graphite-epoxy composites caused by a low-velocity point-mass impactor. The procedures include a dynamic finite element analysis program to calculate transient stresses and strains inside the composite during impact loading and a failure criterion for predicting damage growth. In a more recent attempt by Goldsmith et al. [16], woven carbon-epoxy laminates were subjected to quasi-static as well

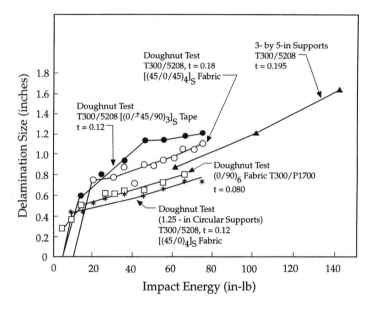

Figure 6.11 Variation of delamination size with impact energy for graphite/epoxy laminates [8]. (Copyright © 1981 AIAA. Reprinted with permission.)

as impact loading, leading to perforation by using cylindro-conical hard steel impactors. The damaged samples were examined to identify various failure modes. A further attempt was made to model major mechanisms of deformation and processes of damage formation that include global plate deflection, fiber breakage, delamination, formation and bending of petals, hole enlargement, and friction between impactor and sample. The conclusions made in this study include observations that the ballistic limit increase correlates approximately linearly with the plate thickness over the range studied for both the Kevlar and graphite laminates, and the perforation resistance for a given thickness of Kevlar laminates is better than that of graphite plates.

6.3 REFERENCES

1. Abrate, S. (1991) "Impact on laminated composite materials," *Appl. Mech. Rev.*, **44**(4), 155–189.

2. Avery, J.G. and Porter, T.R. (1975) *Comparison of the Ballistic Response of Metals and Composites for Military Aircraft Applications*, ASTM STP 568, American Society for Testing and Materials, pp. 3–29.

3. Husman, G.E., Whitney, J.M., and Halpin, J.C. (1975) "Residual strength characterization of laminated composites subjected to impact loading," in *Foreign Object Damage to Composites*, ASTM STP 568, American Society for Testing and Materials, pp. 92–109.

4. Awerbuch, J. and Hahn, H.T. (1976) "Hard object impact damage of metal matrix composites," *J. Comp. Mater.*, **10**, 231–257.

5. Shivakumar, K.N., Elber, W., and Illg, W. (1985) "Prediction of impact force and duration due to low-velocity impact on circular composite laminates", *J. Appl. Mech.*, **52**, 674–680.

6. Greszczuk, L.B. (1982) "Damage in composite materials due to low velocity impact," in *Impact Dynamics*, Editors, J.A. Zukas et al., John Wiley, New York.

7. Sun, C.T. and Chattopadhyay, S. (1975) "Dynamic response of anisotropic plates under initial stress due to impact of a mass", *ASME J. Appl. Mech.*, **42**, 693–698.

8. Dobyns, A.L. (1981) "Analysis of simply-supported plates subjected to static and dynamic loads," *AIAA J.*, **19**(5), 642–650.

9. Goldsmith, W. (1960) *Impact*, Edward Arnold, London, UK.

10. Conway, H.D. and Angew, Z. (1956) "The pressure distribution between two elastic bodies in contact", *J. Mathematics and Physics*, **7**, 460–465.

11. Volmir, A.S. (1967) *A Translation of Flexible Plates and Shells*, AFFDL-TR-66-216.

12. Lukasiewicz, S.A. (1976) "Introduction of concentrated loads in plates and shells", *Progress in Aerospace Science*, **17**, 109–146.

13. Whitney, J.M. and Pagano, N.J. (1970) "Shear deformation in heterogeneous anisotropic plates", *J. Appl. Mech.*, **37**, 1031–1036.

14. Lal, K.M. (1982) "Prediction of residual tensile strength of transversely impacted composite laminates", in *Research in Structural and Solid Mechanics*, Editors, Housner, J.M. and Noor, A.K., NASA CP 2245, pp. 97–112.

15. Choi, Hyuing Yun and Chang, Fu-Kuo (1992) "A model for predicting damage in graphite/epoxy laminated composites resulting from low-velocity point impact", *J. Comp. Mater.*, **26**, 2134–2169.

16. Goldsmith, Werner, Dharan, C.K.H., and Chang, Hui (1995) "Quasi-static and ballistic perforation of carbon fiber laminates", *Int. J. Solids and Structures*, **32**, 89–103.

AUTHOR INDEX

SUBJECT INDEX